XTREME INTERIORS

COURTENAY SMITH + ANNETTE FERRARA

PRESTEL

MUNICH · BERLIN · LONDON · NEW YORK

XTREME INTERIORS
COURTENAY SMITH + ANNETTE FERRARA

PRESTEL
MUNICH BERLIN LONDON NEW YORK

© FOR CONCEPT AND TEXT BY COURTENAY SMITH AND ANNETTE FERRARA 2003
© FOR DESIGN AND LAYOUT BY PRESTEL VERLAG, MUNICH · BERLIN · LONDON · NEW YORK 2003
© FOR ILLUSTRATIONS SEE PHOTO CREDITS, PAGE 159

THE RIGHT OF COURTENAY SMITH AND ANNETTE FERRARA TO BE IDENTIFIED AS AUTHORS OF THIS WORK
HAS BEEN ASSERTED IN ACCORDANCE WITH THE COPYRIGHT, DESIGNS AND PATENTS ACT 1988.

FRONT COVER: HANS HEMMERT, *UNTERWEGS*, 1996
FRONTISPIECE: KRUUNENBERG VAN DER ERVE, *LAMINATA HOUSE*, 2001

PRESTEL VERLAG
KÖNIGINSTRASSE 9, D-80539 MUNICH
TEL. +49 (89) 38 17 09-0
FAX +49 (89) 38 17 09-35
WWW.PRESTEL.DE

PRESTEL PUBLISHING LTD.
4, BLOOMSBURY PLACE, LONDON WC1A 2QA
TEL. +44 (020) 7323 5004
FAX +44 (020) 7636 8004

PRESTEL PUBLISHING
175 FIFTH AVENUE, SUITE 402,
NEW YORK, N.Y. 10010
TEL. +1 (212) 995-2720
FAX +1 (212) 995-2733
WWW.PRESTEL.COM

LIBRARY OF CONGRESS CONTROL NUMBER: 2003107921

THE DEUTSCHE BIBLIOTHEK HOLDS A RECORD OF THIS PUBLICATION IN THE DEUTSCHE NATIONALBIBLIOGRAPHIE;
DETAILED BIBLIOGRAPHICAL DATA CAN BE FOUND UNDER: HTTP://DNB.DDB.DE

PRESTEL BOOKS ARE AVAILABLE WORLDWIDE. PLEASE CONTACT YOUR NEAREST BOOKSELLER
OR ONE OF THE ABOVE ADDRESSES FOR INFORMATION CONCERNING YOUR LOCAL DISTRIBUTOR.

EDITORIAL DIRECTION: PHILIPPA HURD
PICTURE RESEARCH AND MANAGEMENT: SIMONE SCHMICKL
DESIGN, LAYOUT, AND TYPESETTING: SMITH, LONDON
ORIGINATION: DEXTER GRAPHICS, LONDON
PRINTING AND BINDING: PASSAVIA DRUCKSERVICE, PASSAU

PRINTED IN GERMANY ON ACID-FREE PAPER

ISBN 3-7913-2970-7

006 WHAT MAKES TODAY'S INTERIORS SO DIFFERENT, SO EXTREME?
Courtenay Smith and Annette Ferrara

018 CHAPTER 1
SOME ASSEMBLY REQUIRED

022 FORT THUNDER
75, Eagle Street, Providence, Rhode Island

026 IKEA
Furniture and accessories

028 NOTHOZAMILE ZAMA
Zama House

032 SOD HOUSES
Minnesota and Nebraska

034 KEN DRAIZEN
Gas Station House

036 LOT/EK
Miller/Jones Studio

040 RURAL STUDIO
The Lucy House

044 CHARLES STAGG
AV Stagg Art and Wildlife Preserve

048 WORKHOUSE
The Workhouse

050 TSUI DESIGN AND RESEARCH, INC.
Ecological House of the Future

054 CHAPTER 2
GO WITH THE FLOW

058 STEFAN WISCHNEWSKI
Vereinsecke
MTSV Sofa

062 URS HARTMANN & MARCUS WETZEL
Wildbrook

066 SNOWCRASH
Cloud
Soundwave

070 DO-HO SUH
348 West 22nd Street, Apt. A, New York, NY 10011

072 R & SIE...
Barak House

076 NASA
TransHab

080 PRESTON SCOTT COHEN
Torus House

082 ALEKSANDRA KONOPEK
Pneo—A Foldable Habitation Module

084 BERKLINE INC.
Home Theater Seating—
The Cinema Collection

086 MATALI CRASSET
Energizer Room
Phytolab

090 CHAPTER 3
MOVING PICTURES

094 HANS HEMMERT
Nachmittag zuhause in Neukölln
Im Atelier
Unterwegs

098 MARION EICHMANN
16.324.800 Maschen

100 FLORIAN WALLNÖFER, POOL ARCHITEKTUR
T.O.'s Space

104 NATICK LABS (U.S. ARMY)
Force Provider

106 SOFTROOM
Maison Canif

110 KRUUNENBERG VAN DER ERVE
Laminata House

114 ENNEMLAGHI, LTD.
White Apartment

116 RENO DAKOTA
Apartment

120 SEOUNGWON WON
Wunschzimmer—Jörg/Michalis/Zushun/Fabian and Eliane

124 MAURICE AGIS
Dreamspace

126 CHAPTER 4
CONTENTS UNDER PRESSURE

130 FAT (FASHION, ARCHITECTURE, TASTE)
Camo House

132 ATELIER VAN LIESHOUT
Sleep/Study Skull
Tampa Skull

136 DLR GROUP
Jail Cell

AMMAR ELOUEINI, DIGIT-ALL STUDIOS, IN COLLABORATION WITH ARCHEWORKS
Entropias

138 FRANCISCO TORRES
Isolation Room, Psychiatric Hospital of Cery

140 LUCAS MAASSEN
Forever Young: Proportion Conflict

142 ENDEMOL
Big Brother House

144 MASAKI ENDOH & MASAHIRO IKEDA
Natural Ellipse

148 CHRIS BURDEN
Small Skyscraper

150 ABSALON
Cellule no.1 (Réalisation habitable)

152 GREGOR SCHNEIDER
Totes Haus Ur

154 AL V. CORBI, THE DESIGNER
Safe Homes

158 BIBLIOGRAPHY
159 WEB AND EMAIL ADDRESSES AND PHOTO CREDITS
160 ACKNOWLEDGMENTS

WHAT MAKES TODAY'S INTERIORS SO DIFFERENT, SO EXTREME?

...is it paint color? ceiling treatments? oddly shaped windows or avant-garde window dressings? a decorator's or architect's fantasy realized? an accumulation of well-heeled objects? well, yes, and very much no...

CLOCKWISE FROM TOP

COVER OF *MARTHA STEWART LIVING* MAGAZINE.

COVER OF *DWELL* MAGAZINE. AN EXPLOSION OF GLOSSY SHELTER MAGAZINES IN THE 1990S CAUSED MANY A COFFEE TABLE TO GROAN UNDER THEIR COLLECTIVE WEIGHT.

RICHARD HAMILTON, *JUST WHAT IS IT THAT MAKES TODAY'S HOMES SO DIFFERENT, SO APPEALING?*, MIXED MEDIA, 1956. SIGNS OF THE TIME: HAMILTON'S SMALL COLLAGE OFFERED AN IRONIC COMMENTARY ON THE POST-WORLD WAR II MODERN INTERIOR.

In 1956, British artist Richard Hamilton signaled a new era in art history and by default interior design with a small but iconographically dense collage he exhibited in the seminal Independent Group exhibition "This is Tomorrow" at the Whitechapel Art Gallery in London. Titled *Just what is it that makes today's homes so different, so appealing?*, Hamilton's tiny pastiche offered a ribald rendition of the new modern interior, with all its modern wonders and up-to-date styling. In addition to Danish furniture and drip-print carpet (strikingly similar to a Jackson Pollock painting), Hamilton threw in some canned ham, a reel-to-reel tape recorder, a vacuum cleaner, a naked muscle man holding a Tootsie Pop, comic book covers presented as framed keepsakes, and an Al Jolson poster. Hovering above it all is an image of Mars, a harbinger of the future soon to come.

An artwork that many consider to be the defining image of what we now call Pop Art, Hamilton's pastiche presented the domestic interior as a "realm of a private fantasy identified with the very public fetishism of mass consumption"[1] and earmarked narcissism, ephemerality, mobility, and transportability as defining concepts—good or bad—of the time. Nearly 50 years later, the issues raised by Hamilton's provocative collage continue to be relevant in our post-post-modern society, which is why we have appropriated its title (and tweaked it a bit) to ask anew what it is, exactly, that defines the domestic dwellings of *our* time.

Just what is it that makes today's interiors so different, so extreme? Is it paint color? Ceiling treatments? Oddly shaped windows or avant-garde window dressings? A decorator's or architect's fantasy realized? An accumulation of well-heeled objects? Well, yes, and very much no. Sometimes anything out of the ordinary can, in certain circumstances, register as "extreme." But what do interiors that go far-far-left-of-center look like—those that don't appear on most cultural radars, that, maybe, wouldn't even qualify as interiors to some? It is the quest of this book to answer these questions and, perhaps, if we've done our job, to provoke even more.

To begin with, we're not talking about the art of feng shui, even less about interior design orchestrated by someone who has come in to re-cover the furniture, coordinate duvets and drapes, and paper the walls. We're also not concerned with "interior design" *per se* but rather with questions such as "what's it like to be alive in the twenty-first century?," "what spaces are best suited to that existence?," and "what is an interior in the first place?" Certainly, the decorating profession is still going strong, but we are more interested in the numerous global forces that have reshaped the interior altogether, not just changed its appearance. As demonstrated by the explosion of magazines with titles such as *Shelter*, *Nest*, *Dwell*, *Surface*, *Frame*, and *Martha Stewart Living* glossy decorator mags are now sharing shelf space with monthlies and quarterlies devoted to a new kind of space-making and the debate has shifted from, say, smooth versus textured walls to whether to have walls at all. We, too, are interested in defining what a domestic interior *is* beyond the role of professional interior design by identifying some of the influences that have come to bear on it.

To begin with, the words shelter, home, or refuge spring immediately to mind, and, of course, family. Domestic environments provide us with comfort, protection, a sense of security. They are a demilitarized zone of sorts—a physical and mental demarcation of where "us" ends and "them" begins, with enough plasterboard, wood, and high-tech security systems in between to keep "them" out. They provide us with a space to relax, to be accepted, to be our real selves. Traditionally, a home inspires in us a sense of longevity or invokes a sense of nostalgia. Homes are the stuff of childhood dreams and adult longing, their rooms the stage sets of oedipal dramas and domestic squabbles as well as the backdrops of millions of "Kodak moments." Domestic interiors and their attendant furnishings are an expression of self, a reflection of class and social standing, and sometimes a social space where we entertain friends and family in addition to entertaining ideas about our economic caste. They are a reflection of an inhabitant's or a family's values, and tell a powerful visual history loaded in iconographic detail. Sometimes the work products of architects or designers, domestic interiors can be another entry in the professional's catalogue raisonné.

Feminists have taught us to look at domestic interiors as gendered spaces. Interior = softness = female. Exterior = hardness = male. Homemaker versus the home builder and the long, drawn-out antagonism between the interior designer (traditionally, female) and architect (many times, male). Domestic spaces can be instructive, teaching girls how to be women and boys, men. Societal preparation for girls takes place in the nurturing confines of the kitchen/dining room and the cleansing spaces of the bath and laundry. Boys retreat to the home office (formerly the reclusive or learning spaces called the library or den) and the garage or basement, when they're not outside, that is. Family life, together, is acted out in the communal area of the living room and around the dining table. Formal, or social life, is played out in the salon or front room.

After thoroughly parsing the notion of "domestic interior," we set off to extrapolate the meaning of extreme (or "xtreme," which seems a degree more) and its relation to the domestic interior. Combing through hundreds of possible projects, it became clear that it wasn't just one thing alone that made an

WHAT MAKES TODAY'S INTERIORS SO DIFFERENT, SO EXTREME? COURTENAY SMITH AND ANNETTE FERRARA

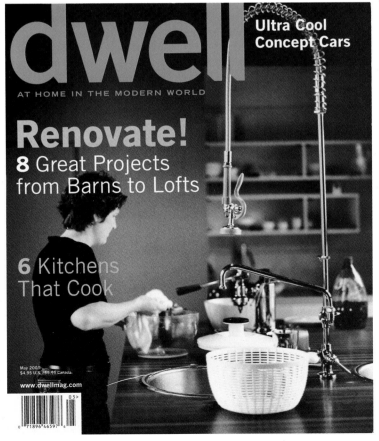

RIGHT TOP BARBARA GALLUCCI, *MARILYN (HOLE REVEALED) LEVITTOWN INTERIOR, 1950 RANCH HOUSE*, 2002. C-PRINT, 16 x 20 INCHES. LEVITTOWN, LONG ISLAND, JUST OUTSIDE NEW YORK CITY, IS RENOWNED FOR BEING THE FIRST PLANNED COMMUNITY OF MASS-PRODUCED, SINGLE-FAMILY HOMES IN AMERICA. IN THE 1950 RANCH HOUSE THE TELEVISION PARTS WERE SITE-SPECIFICALLY INSTALLED RIGHT INTO THE WALL AND THE HOUSE ITSELF BECAME ITS CABINET.

RIGHT BOTTOM MICHAEL SAILSTORFER (WITH ALFRED KURZ), *WALDPUTZ (FOREST CLEANING)*, 2000. THROUGH A SIMPLE ACT OF SQUARING OFF AND CLEARING A SPACE IN THE WOODS, ARTISTS SAILSTORFER AND KURZ CREATED AN ARTIFICIAL "ROOM" WITHIN A NATURAL ONE.

what makes today's interiors so different, so extreme? courtenay smith and Annette Ferrara

interior extreme but a culmination of factors. First off, some extreme interiors seem to challenge notions of comfort and protection, either by eradicating these functions altogether so that either ephemerality and fragility dominate—think of the spartan, transportable accommodations of a mountain climber—or by expanding them exponentially so that permanence and impenetrability take hold as in prisons. some extreme interiors, like Liberace's ultra-glitzy outfits and Marilyn Manson's monstrous makeup, defy reigning notions of "good" taste, and, occasionally, the architect / designer / inhabitant's aesthetic is so unique or single-minded that not only is there no precedent for it, but it can also never be reproduced.

Sometimes a set of extreme circumstances (geographical, political, economic) force interior space to be born spontaneously. In times of war or during certain desperate situations, anyplace can turn into a home. Suddenly tent systems, haphazard accumulations of cardboard, or various buildings provided by disaster relief organizations, are welcome homes away from home for those displaced. The result of dire economic necessity, we see homes around the globe cobbled together with scraps of metal, aluminum cans, or industrial leftovers, their ersatz shapes and tiny rooms containing the barest of furnishings that have likewise been pilfered or rescued from the dustbins of the more fortunate. In the western hemisphere, consumer frenzy has certainly intensified manifold, resulting in, among other things, landfills, pollution, global warming, war, chemical sensitivity disorders, and a widening gap between the "haves" and the "have-nots." Paradoxically, it is the ever-increasing array of products that allow us to define ourselves the way we like, to personalize, realize, and materialize the spaces we inhabit. we are spoiled by these choices and we, in turn, despoil the land through their use and disposal. we have found an ingenious few, though, who have done their best to create interiors by recycling these post-industrial remnants or by attempting to live off the grid in radically "green" or ecological homes.

A number of extreme spaces we looked at were born from the development of new technology. computer-generated architectural forms have granted architects the freedom to explore the curvaceous possibilities of the "blob," much to the chagrin of devotees of the "box." when walls melt into floors and then become ceilings, the movement of individuals and information within that space is changed in intriguing ways. other types of technology allow some—mostly soldiers and other professionals of their ilk—to live in formerly uninhabitable places such as outer space, underwater, or in other unlikely environments. For these intrepid individuals, because their *raison d'être* is extreme, their domestic surroundings naturally follow suit. In other cases, the desire to utilize unusual or new building materials such as latex, laminated glass, PVC, nylon, or resin, have resulted in unique residences, and similarly the wish to challenge conventional ideas of space-use in a radical way has produced projects such as an infinitely transformable, swiss Army knife-like room system.

Many projects we looked at seemed to be extreme because they crossed the boundaries between art/life/architecture/design. As professional art historians (full disclosure, here), we naturally gravitated to examples of contemporary artists who blur the distinctions between these formerly distinct practices. contemporary artists interested in installation art and its attendant possibilities have been generously borrowing ideas from avant-garde architects and designers by incorporating references to architectural and design histories into their art work, collaborating with practicing architects to create interiors and structures, or by creatively exploring ideas for mutable furnishings and mobile housing.

In the end, we held to a criteria of domestic interiors, choosing to focus on the places where we spend most of our time rather than on other projects that might also be deemed as extreme, such as mini storage shelters, theme parks, and clubs. And yet we were catholic in our selection of whom we felt should be included as designers of an extreme interior, since the projects of renegade professional architects or interior designers are often the only ones examined in such a book. we looked to creative individuals, collectives, corporations, artists, the U.S. Armed Forces, and NASA—in addition to the professionals—to see what types of interiors they were producing and how these were affecting the spaces we call home. By scanning the output of this heterogeneous group of "designers," patterns of ideas started to form. We grouped similar interiors together to make sense of the diverse concepts and allow them to be read as cultural tea leaves. The individual chapters represent the dominant concerns—the defining themes—of our times in terms of domestic interior design. They are a snapshot of the *zeitgeist*.

In the chapter "Some Assembly Required," for instance, a nostalgic longing for the handmade seems prevalent as western culture runs headlong into the virtual worlds of information technology. Projects covered in this chapter evidence a desire to "do-it-yourself," to create a home that is unique, from the bottom up and the top down, and with unconventional materials. Or at least to customize it with the help of IKEA, a store that commissions goods from prominent designers, providing eager consumers with flexible, well-designed, and affordable furnishings for their space-starved dwellings. To contemporary urban hipsters, low on cash but learned in the

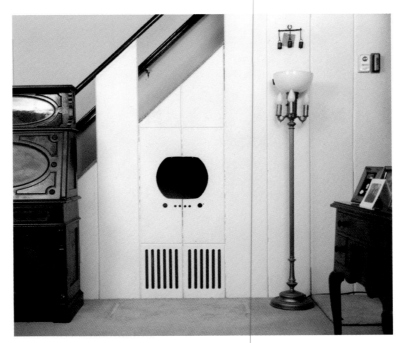

CLOCKWISE FROM TOP LEFT

SHIGERU BAN'S *NAKED HOUSE*, 2001, FEATURES ROOMS THAT ROLL.

JUST BECAUSE YOU'RE MOBILE DOESN'T MEAN YOUR PERSONAL HYGIENE HAS TO SUFFER—MICHEL PERTHU'S COLLAPSIBLE SHOWER UNIT, 2002.

IN CENTRAL MONGOLIA, NOMADIC SHEPHERDING TRIBES CALL THEIR TRADITIONAL GER TENTS HOME.

FOR THE HIGH-FLYING EXECUTIVE, ONLY THE (SUBURBAN) HOME-OFFICE COMFORTS OF THE *YAK-42 FLYING OFFICE* WILL DO.

realms of contemporary design, cheap is not only good but is a requirement. within the last few years there has also been an increasing interest in the handmade as evidenced by the growing communities reading getcrafty.com and *Readymade* magazine, and participating in the activities of The church of craft. cost-effective and spiritually rewarding, DIY can also be a political backlash against consumerist, homogeneous culture. let's face it, suburban houses with their Dryvit facades just don't turn true DIY-ers on. Detritus, decay, and the products of throwaway culture gone amok seem to provide the building materials that the intrepid do-it-yourselfer craves, maybe because these materials were once loved, or have a sense of time imprinted on them—qualities that cannot be found in the sleek metallic/plastic realms of technology. whatever motivates this trend toward the reclamation of the remnants of industrial society, it can be seen as a form of innovative recycling and the product of creative frugality at its best.

Another key motivator is a desire for mobility or, if that term is abstracted, "informational flow," as discussed in the chapter entitled "Go with the Flow." Attentive readers of stylish periodicals will recognize a term that has been bandied about in them for a couple of years: "global nomads." Like the "original" nomads, tribes who moved restlessly in order to maintain their hunting and gathering lifestyles, global nomads expect their homes to adapt to any situation. The products of the boom economy of the 1990s which emphasized on-the-go entrepreneurship, global nomads have benefited from the rise of the Internet and other web-based technologies as well as the increasing affordability of international travel. They, along with most of the rest of us, are now so accustomed to the "non-space" of informational space that mobility, flexibility, and the ability to communicate within that space endlessly, wherever they are, is a necessity for them. At the height of the new Network Economy, products such as the luxurious YAK-42 *Flying office* were created in response to these demands, giving extra meaning to the now quaint term, "home/office."

Members of our generation often move away from their birthplaces for school, then on to new cities for jobs, and then again to greener fields until "home" no longer means the gathering place of their family of origin (heck, mom and dad probably moved several times, too), but denotes the place where we store our iMacs when not in use. This desire to be on the go while still retaining a sense of (semi-)permanent home is reflected in the interiors produced by individuals, artists, architects, companies, and designers alike who create movable rooms, collapsible fixtures, multi-functional dwellings, and double-duty furnishings. Their "places" often tend to utilize computer-aided design strategies and transparent, movable materials so that they can be transformed almost effortlessly to accommodate different needs and desires. In Japanese architect shigeru Ban's *case study House 10E*, for example, permanent walls are gone, replaced with portable, box-like rooms on wheels floating in a giant, open-plan space. The boxes serve as sleeping quarters, living rooms, studies—whatever is demanded of them from moment to moment. Traditionalists beware: a kitchen isn't just a kitchen anymore, and, careful, you might mistake it for the bathroom or the office.

yet another factor that cannot be overlooked as an influence on today's domestic interior is the seamless integration of visual media, in all its forms, into our daily lives. Image is everything and pictures—on the TV, on our cell phones, on our computers, on billboards, in the cinema, and across the pages of magazines—play a formative role in how we imagine and construct our habitats and ourselves. Images conjure desire and we often tend to mimic what we see, a theme that we discuss in the chapter called "Moving Pictures." In earlier times, lords and kings employed artists to reproduce their grand halls and earthly possessions in paintings of architectural scale. These theatrical, carefully composed "grand views" recorded, in an encyclopedic way, the wealth and good taste of the client and illustrated to others the span of his realm and power. Nowadays, photographic and digital images have usurped paint on canvas as a preferred means of reproduction but they continue to play a role in documenting our living spaces and our stuff. The digital camera in particular has made our perception of the world we inhabit quick, mobile, and disposable. what we see is not necessarily what we must accept and with the help of a delete button, we can eliminate views we don't like and re-shoot until we do. with the help of a computer, we are also able to shape our world in as many ways as there are to manipulate a digital image, and if images of our living spaces can be cut, pasted, enlarged, reduced, morphed, collaged, and deconstructed, then why not the spaces themselves? Here we can also not ignore the enormous influence of the film industry in shaping our collective experience of interiors. The ability of film architecture to simultaneously mimic reality and invoke fantasies and dreams has made it a fertile testing ground for new ideas about space.

Finally, control and discipline are also two factors that come to bear on our notions of space in the last chapter, "contents under Pressure." A paradox of the contemporary "private" interior is that in certain situations its design forces the behavioral expectations of others on to us. Prison cells and mental wards are the most obvious cases in point, where surveillance, combined with tiny living quarters and strict regulations about personal

what makes today's Interiors so Different, so Extreme? courtenay smith and Annette Ferrara

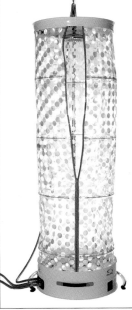

CLOCKWISE FROM TOP LEFT IN *THE HOUSE OF RABBI LÖW* FROM PAUL WEGENER & CARL BOESE'S *THE GOLEM: HOW IT CAME INTO THE WORLD*, PROJEKTIONS-A.G. UNION, GERMANY, 1920. EUROPEAN FILM SETS OF THE EARLY 20TH CENTURY WERE FERTILE TESTING GROUNDS FOR NEW IDEAS ABOUT SPACE AND OFTEN PRESENTED THE INTERIOR AS A SURREAL, EMOTIVE, EVEN SCULPTURAL EXPERIENCE.

DURING THE MID-20TH CENTURY, INVENTIVE ARCHITECTS SUCH AS BRUCE GOFF BEGAN INVESTIGATING THE DYNAMICS OF ALTERNATE GEOMETRIES. HERE *TRIAERO*, COMPLETED IN 1942, A CANTILEVERED, TRIANGULAR-SHAPED GLASS HOME INFLUENCED BY GOFF'S LONGTIME FRIEND, FRANK LLOYD WRIGHT.

PAUL KLEE, *MASTER'S HOUSE*, DESSAU, GERMANY, 1926. ONE OF A TRIO OF HOUSES DESIGNED FOR BAUHAUS FACULTY BY WALTER GROPIUS, THE KLEE HOUSE CONTINUES TO BE A LANDMARK EXAMPLE OF AN INTERIOR EXPRESSED IN THE LANGUAGE OF ABSTRACT PAINTING, WITH EACH ROOM AND WALL DEFINED BY BRIGHTLY COLORED PLANES OR BLOCKS OF COLOR.

ROBERT BRUNO, *STEEL HOUSE*, LUBBOCK, TEXAS, USA, 1978–2002. THE EXPRESSION-ISTIC, CATHEDRAL-LIKE INTERIOR OF BRUNO'S SELF-BUILT HOME DEMONSTRATES A CONTEMPORARY TENDENCY TOWARD PERSONAL FANTASY.

KURT SCHWITTERS, *MERZBAU*, HANNOVER, GERMANY, 1919–37. THE GERMAN ARTIST'S MAJOR WORK, FIRST BEGUN IN HIS STUDIO IN HANOVER, WAS A RAMBLING AND EVER-CHANGING AGGLOMERATION OF ARCHITECTURAL ELEMENTS AND FOUND OBJECTS WITH NO DEFINED BORDERS OR LIMITS.

what makes today's interiors so different, so extreme? courtenay smith and Annette Ferrara

possessions, subjugate the individual's will to that of the group. But what happens when interiors such as these begin to cross over into our private dwellings? Paranoia and xenophobia are paving the way for domestic designs that imprison "free" inhabitants in alarmed paradises in an effort to keep undesirables at bay. The impact of the September 11, 2001 terrorist attacks cannot be ignored here, and the aftermath still presents itself in an overwhelming glut of safe rooms and home security systems. What modern interior would be complete without a kevlar bed that resists a missile attack; a panic room that doubles as a haven against chemical warfare; and in-home security cameras to make sure the babysitter doesn't harm the kids?

The man-made environment also holds sway over "interior design" by producing toxins and chemicals that drive some individuals into unwanted exile within anti-allergy rooms or sanitized trailers. Multiple chemical sensitivity—made famous in Todd Haynes' 1995 film *Safe*—and Gulf War chronic fatigue syndrome are but two of a growing number of severe allergies to synthetic products, which confine individuals to cells like prisoners and force them to readjust their daily habits (in order to avoid substances that will harm them) with the patience and discipline of a monk.

Likewise, self-discipline is a factor that plays a role in interior design, especially in dwellings by individuals who choose to retreat entirely into worlds of their own making. Artists are the forerunners here and have come up with spaces that deprive the senses, function as retreats, distort scale, or compress space in unnatural ways.

But let's step back a moment and ask, "How did we get here?" The extreme interiors pictured in this book didn't arrive *sui generis*, with light-filled PVC chambers and kevlar beds intact. No, it was a longer process and one that bears its share of responsibility for why these interiors look the way they do today, at the beginning of a new century. A look back over the last century of landmarks in architectural and interior design will provide a much-needed foundation for these interiors.

In the early part of the century, interiors were heavily influenced by the intersecting activities of artists, architects, and filmmakers, especially those working in Germany. Recovering from the atrocities of World War I— a war, by the way, that ended monarchies and gilded interiors at the same time—and eager to move into a proletariat-led modern era, filmmakers such as Paul Wegener capitalized on the possibilities of the medium to experiment with new and highly expressionistic forms. For his 1920 film, *Der Golem*, set in a sixteenth-century Prague ghetto, Wegener employed architect Hans Poelzig to design the sets. Rather than recreating a historical verisimilitude, Poelzig built a non-hierarchical suggestion of a Jewish ghetto out of sculptural forms that disregarded structural considerations and that were veiled in shadow. Indeed, the cave-like interior and orgiastic, anthropoid forms of his *Rabbi Löwe's House* prefigure visionary interiors to come.

At roughly the same moment, Dada artist Kurt Schwitters began to erect a real-life expressionistic interior in his studio in Hanover. Referred to as the *Merzbau* (1919 – 1937), the continuously changing assemblage was a three-dimensional collage of found materials, objects stolen from friends, and sculptural forms affixed to the existing architectural structure, and later emerging from holes he cut into the ceiling and floor. Like his latter-day DIY offspring, Schwitters found visual poetry in the cast-off junk of modern society and used the detritus as the core of his metabolizing *gesamtkunswerk*, an all-encompassing environment with no discernible beginning or end.

Running parallel to these subjective tendencies were the objective structures of architects such as Walter Gropius at the Bauhaus in Dessau. His *Master's Houses* of 1925 were solid statements in order and geometry. Hard-edged cubes topped with flat roofs and framed by large windows, each machine for living was marked by a rational floor plan with square rooms divided by function—which would become a hallmark of building after World War II. Inside, the rooms were painted in lively hues that turned the interiors into abstract, flat planes of shifting color, light, and shadow.

In the middle part of the century, houses started changing shape. Suddenly, the square plan of many post–World War II homes was, well, "square," and adventurous architects began looking to other geometrical compositions to get their fix. In America in the 1940s, Bruce Goff, influenced by longtime friend Frank Lloyd Wright, designed *Triaero*, a radical, cantilevered, low-lying triangular-shaped glass structure. Goff's design was inspired by a natural clearing that was triangular in shape, but his thinking about the relationship between the house and its site and interior/exterior dynamic didn't stop at its glass, copper, and redwood cladding. Like Wright, he custom-designed all of the furnishings, including a triangular dining table and angled storage units. Interior design was not relegated to a separate (or female) profession or homeowner but was deemed worthy enough to deserve the full attention of the architect. The house was not just a decorative container for its discordant furnishing, but it also became desirable for the inside and outside of the house to be in aesthetic harmony.

Two decades later, John Lautner's *chemosphere*, a space-age, ultra-hip dwelling, sprang up on the Left Coast. Forget low-lying

CLOCKWISE FROM TOP LEFT

ANDY WACHOWSKI THE MATRIX, 1999. THE FILM OPENED UP MYRIAD POSSIBILITIES FOR THINKING ABOUT INTERIORS, IN PARTICULAR, THE IDEA THAT THE WORLD WE INHABIT IS SIMPLY A PROJECTED IMAGE WITHIN A WHITE VOID—A SPACE THAT MERELY NEEDS TO BE FILLED TO BE REALIZED.

PAINTER, SCULPTOR, ARCHITECT, AND LANDSCAPER CÉSAR MANRIQUE WAS NOT CONTENT TO BUILD HIS DREAM HOME, TARO DE TAHÍCHE, ABOVE GROUND, BUT INSISTED ON CREATING STAR TREK-LIKE ROOMS OUT OF VOLCANIC BUBBLES BURIED DEEP IN THE CANARY ISLAANDS.

VERNER PANTON, VISIONA 2, COLOGNE FURNITURE FAIR, 1970. THE SWINGING SIXTIES WOULDN'T HAVE BEEN AS GROOVY WITHOUT VERNER PANTON'S FLOOR-TO-CEILING PSYCHEDELIC INTERIORS.

QUASAR (NGUYEN MANH KHANH), CYLINDRICAL INFLATABLE HOUSE, 1968. WITH THE ADVENT OF THE S PACE AGE, INFLATABLE ENVIRONMENTS GAVE THE IMPRESSION OF INFINITE MOVEMENT IN SPACE.

WARREN CHALK, PETER COOK, DENNIS CROMPTON, RON HERRON, 1990 HOUSE, 1967. A FULLY AUTOMATED USER-CONTROLLED ENVIRONMENT IN WHICH WALLS, CEILINGS, AND FLOORS COULD BE CHANGED ACCORDING TO THE INHABITANTS' DESIRES.

STANLEY KUBRICK, 2001: A SPACE ODYSSEY, 1968. A TIMELESS, PLACELESS, PLACE AND THE CULMINATION OF ALL INTERIORS PAST, PRESENT, AND FUTURE.

---

ranch houses, the Hollywood Hills sported a house on a pole that looked more like a UFO or air traffic control center—a perfect abode for freshly minted jetsetters. The interior dynamic of the house, now stripped of its beloved ninety-degree angles, was now beholden to a modified centrifugal force. A clear span of over 1,300 square feet of living space also created greater flow through the house, allowing inhabitants and guests to converse easily as well as reconfigure the space in ingenious new ways. Today's loft living is much beholden to Lautner's dissolution of conventional room divisions.

As Lautner and Goff were experimenting with ground-level architecture, césar Manrique—painter, sculptor, architect, landscaper—was burrowing deep below the Canary Islands to build his dream home, Taro de Tahíche. His unusual house straddles massive sheets of volcanic lava above ground as well as delves hundreds of feet into subterranean chambers left after eruptions. The upper half of the house is late 1960s modernism-as-usual, but below, Manrique designed color-coded rooms—red, black, yellow, white, and avocado—out of interconnecting ovoid "bubbles" or lava cavities. The "architecture" of the interior space was created naturally, by chance. With Manrique's eye for 1960s' Italian loucheness, though, the bubbles were turned into Star Trek-like party spaces for the designer and his mod friends.

An insistence on a totality of vision in interior design is a recurring theme in the twentieth century. The bad boy of post-war Dutch design, Verner Panton, created groovy, floor-to-ceiling and back-down-the-walls-to-the-floor, psychedelic interiors that proved the perfect backdrop for similarly attired occupants. Interior as tableaux, Panton created total atmospheric experiences, pushing the design envelope as far as it would go and causing a great deal of disorientation in the process.

The British architectural collective, Archigram, also subscribed to an all-encompassing view of interior design and were greatly influenced by the technological possibilities that the space age ushered in. Motion, flexibility, self-sufficiency, and independence through automation were heavily stressed in Archigram's designs for living and their prototype 1990 House of 1967 was a user-controlled environment in which walls, ceilings, and floors were "conditions" that could be changed according to the inhabitants' desires. A chair-car, moveable robots, and wide-screen television with surround-sound were also among the appliances available at the touch of a button.

The space age also popularized inflatable furniture and enclosures that contributed to a sense of carry-and-go freedom and gave the impression of infinite movement within space. At the same time, man's foray towards the final frontier shifted our understanding of infinite space from a theoretical construct to fuzzy pictures of the outer limits produced by NASA. The newfound fascination with bending, unending, open-ended space was nowhere more apparent than in the last scenes of Stanley Kubricks's classic 2001: A Space Odyssey (1968) which were staged in a cool white bedroom amid a surreal arrangement of rococo-esque furniture, classical statuary, and backlit floor tiles. Vaguely familiar but ultimately placeless, the room seems to represent the sum of all interiors past, present, and future. The idea that an interior could exist in another dimension for all time (and perhaps even without our knowledge) was later taken up in the film The Matrix (1999), whose main characters periodically visit the "matrix," a false machine-generated reality that appears as an infinite white zone waiting to be filled.

But let's get back to our third-dimensional reality and the task at hand. We started our inquiry into what makes today's interiors so extreme with the help of Richard Hamilton's pop collage of the 1950s. It seems fitting that we try to conjure our own contemporary picture of what we have discovered during our "xtreme" journey into interiors.

It's likely that our "collage" of the contemporary state of interior affairs would be digital and be projected on a scrim instead of hanging on a wall. Hamilton's Tootsie Pop-holding strongman would be replaced with a woman adorned with a Home Depot toolbelt, sitting on an IKEA stool, accessing her Microstation program to reconfigure her home's layout. The area in which she sits would be a large, lofted space filled with inflatable rooms, furnished with lounging stations made of recycled backpacks, and containing a kitchen/bathroom unit that doubles as a sculpture. What walls there are would be made of a combination of laminated glass, coca-cola cans, and latex. Outside one of the portholes, you would catch a glimpse of the surrounding underwater life.

Or perhaps our "collage" would be more theoretical: an accumulation of points in space, connected together only in our imagination, where the boundary between interior and exterior is simply not applicable, its mass made up of random molecules of oxygen strung together in an intricate, invisible pattern.

---

[1] Thomas Lawson, "Bunk: Eduardo Paolozzi and the Legacy of the Independent Group," *Modern Dreams: The Rise and Fall of Pop* (Cambridge/London: MIT, 1988), p. 25.

what makes today's interiors so different, so extreme? courtenay smith and Annette Ferrara

CHAPTER 1

SOME ASSEMBLY REQUIRED

Pity the poor soul who can't hang wallpaper or install a garbage disposal, for these are the halcyon days of the DO-IT-YOURSELFER. Steadfastly refusing decorating advice, preferring his or her own homegrown aesthetic to the bland ideas of the professional designer, the DIYer knows how to turn garbage into gold.

DIYers are easy to spot. Their coffee tables (if they have chosen to build such bourgeois items) are laden with IKEA catalogues and subscriptions to Nest, ReadyMade, and Martha Stewart Living. They religiously tape episodes of This Old House and Changing Rooms and are card-carrying members of The Church of Craft. They are able recite, from memory, recipes for more than three types of cement. And they will always choose a trip to the junkyard or Home Depot to that of a furniture store.

      In this chapter, we celebrate the gusto, determination, and creativity of DIYers and their obsession with recycling what others foolishly deem trash. What unites this disparate group that includes architectural stars LOT/EK, Eugene Tsui, and Rural Studio, the Fort Thunder collective and The Workhouse (created by "Peter"), and a number of individual homeowners, is a proclivity towards thriftiness in home interior design; a single-mindedness or uniqueness of design vision; a willingness and desire to work with humble or cast-aside materials; and an approach to interior design that is intuitive and organic.

      In the 19th century, when intrepid pioneers moved west across the United States to claim their homesteads, sod houses were the ultimate DIY homes. As they approached the Midwest with its dearth of trees, these pioneers, with their American ingenuity, looked to the dense prairie sod beneath their feet to provide them with cozy shelters. Today, sod homes have become the ultimate statements in "green' architecture. If the thought of living "underground" while above ground doesn't suit your more modern tastes, then head to IKEA, the Swedish company who has taught us all that chic is not incompatible with cheap. IKEA has emboldened millions of us to forgo the costly fees of professional interior decorators and pursue the Nordic dream of self-sufficiency, good design, and affordability.

      The New York-based architectural group LOT/EK has certainly been inspired by the scrappiness of the DIYer. LOT/EK has made its name by thoughtfully reinventing uses for cast-aside industrial objects such as freight containers in their home designs, and reinvestigating the potential of durable, cheap, and readily available building materials. Likewise, architect Eugene Tsui pushes the boundaries of recycling in his Ecological House of the Future, a house that cleans and reuses its gray and black water with the help of its own reconstructed wetlands.

The driving need for individual self-expression and an intense dislike of waste unites the efforts of many in this chapter. The young artists and musicians who made up the recently disbanded Fort Thunder commune in Rhode Island turned to the offerings of urban dumpsters and their own unbridled creativity to transform their sparse industrial space into a unique, riotously decorated home. Traditional homes and apartments with their cookie-cutter aesthetics never appealed to Ken Draizen or Charles Stagg, so when they redecorated their apartment or were able to construct their own homes, they took home design into their own hands, literally. The outcomes: gas station-house, and the AV Stagg Art and Wildlife Preserve.

For Nothozamile Zama, The Workhouse, and the architectural students who participate in Rural Studio, Samuel Mockabee's revolutionary humanitarian architectural program, home improvement is closely tied to self-improvement and human dignity. A home, while a reflection of the self, also has the potential to transform and inspire those who create it and those who reside in it. In the hands of Mrs. Zama, "Peter", and Rural Studio, interior design rises above mere decoration or faddish style to incorporate economic necessity, compassion, and a will to survive.

1-22 FORT THUNDER

75 Eagle Street, Providence, Rhode Island
1995, demolished 2001

RIGHT BRIAN CHIPPENDALE'S SPACE IN FORT THUNDER WAS COMPRISED OF TWO LARGE SPACES DIVIDED BY A PLYWOOD WALL COVERED WITH SCREEN-PRINTED WALLPAPER. THE TRAPEZOIDAL DOORWAY LEADS INTO A MUSIC REHEARSAL SPACE.
THE SPACE WAS DEEMED "FORT THUNDER" BECAUSE ITS INHABITANTS WERE ALLOWED TO BE AS NOISY AS THEY LIKED 24/7.

OPPOSITE, LEFT A VIEW INTO THE KITCHEN. WHAT'S DETRITUS AND WHAT'S DECORATION? ONLY THE OCCUPANTS COULD TELL FOR SURE.

OPPOSITE, RIGHT TOP THE SCREEN-PRINTING STUDIO WHERE COMIC BOOKS, RECORD COVERS, AND POSTERS FOR LOCAL PROVIDENCE EVENTS WERE CREATED.

OPPOSITE, RIGHT BOTTOM PERHAPS THE MOST RECOGNIZABLE ROOM IN FORT THUNDER: THE LIBRARY.

Anyone who has visited punk rock clubs knows how detritus and decay can become house style. But how many people can say they've lived in settings where, from floor to ceiling and back again, post-consumer waste—broken toys, old Christmas decorations, stuffed animals, posters, stickers—becomes artful décor? Founded by four Rhode Island School of Design students, the riotously decorated Fort Thunder, demolished in 2001 to make room for a neighborhood revitalization project, was the home of the Providence noise/indy rock scene. Colonized in 1995, the 19th-century industrial space was home to 21 men and four women during its short life span. How did Fort Thunder's decorating scheme manifest itself? "In the beginning," explains Jim Drain, a former resident, "it was all white. Then everyone started drawing on the walls, and then suddenly everything was covered. It was like a disease." In this freewheeling space, design was dictated by whimsy—"we didn't really intellectualize a lot of stuff. Everything that happened in the house we kind of let happen," explains former Fort co-founder Brian Chippendale. Eschewing a logical floor plan, the space was haphazardly divided into common areas such as the kitchen, band room, and screen-printing annex, and then private spaces, which doubled as individual studios and living areas. When a new resident moved in, walls were simply erected or torn down to create new private rooms.

Fort Thunder, , Providence, Rhode Island

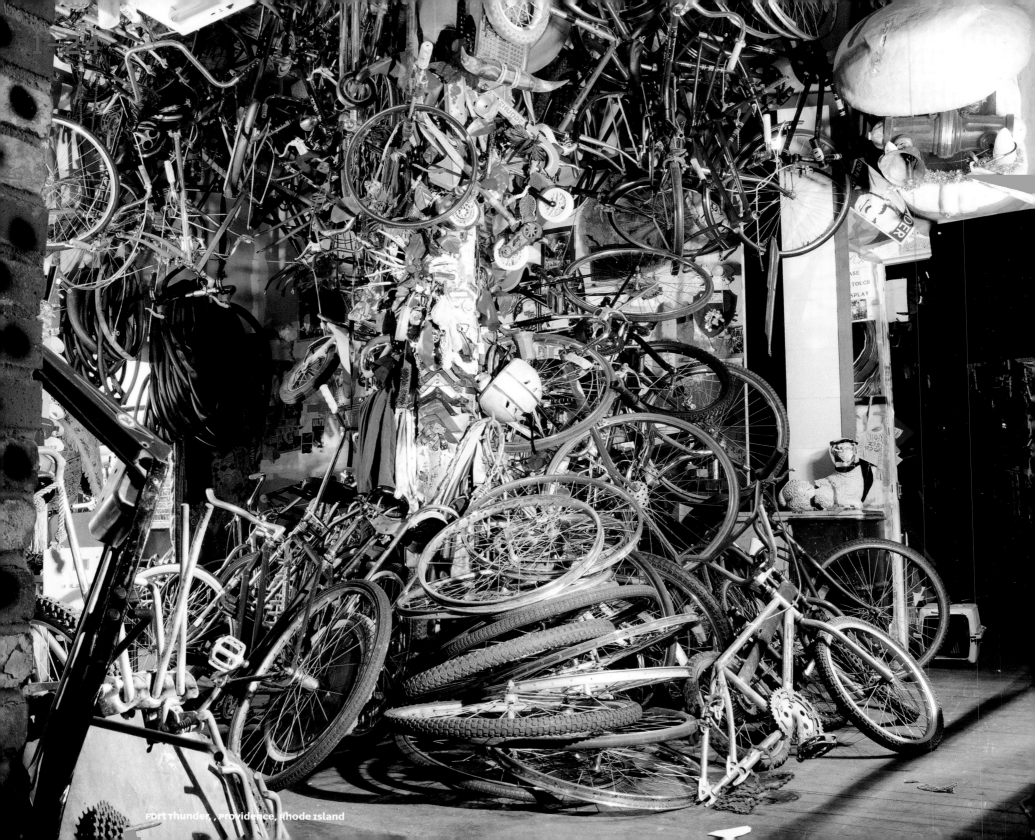

Fort Thunder, Providence, Rhode Island

LEFT FORMER RESIDENT PETER FULLER WELDED PARTS OF OLD BIKES TOGETHER IN HIS SPACE TO CREATE "CHOPPERS," OR UNUSUAL LOWRIDERS.

RIGHT IN FORT THUNDER'S TOPSY-TURVY WORLD, UPSIDE DOWN IS A-OKAY. HERE, A COMMUNE MEMBER SUSPENDS HIMSELF FROM ONE OF THE MANY PEEPHOLES DOTTING THE SPACE. A SHALLOW ATTIC SLEEPING COMPARTMENT WAS ABOVE THE STUFFED-ANIMAL CEILING.

1–26 IKEA

sweden, 1954 – present

ABOVE AND OPPOSITE THE 2003 EDITION OF IKEA'S SALES CATALOGUE—DISTRIBUTED TO OVER 118 MILLION CUSTOMERS IN 23 DIFFERENT LANGUAGES—ENCOURAGES TODAYS DIYERS TO "LIVE HIGH" AND "DISCOVER MORE SPACE" BY THINKING OF EVERY SURFACE IN THEIR INTERIORS AS USEABLE SPACE. A CARTOON "BOX" AND BIRDS-EYE VIEW ILLUSTRATE HOW WALLS CAN BECOME MORE THAN JUST A PLACE FOR PICTURES. SHELVING THAT RISES TO THE OCCASION, CHAIRS THAT HANG WHEN NOT IN USE. THESE AND MORE OPTIONS ARE HIGHLIGHTED IN MORE THAN 375 PAGES OF PRODUCTS AND SOLUTIONS THAT SQUEEZE MORE SPACE OUT OF ANY ROOM, REGARDLESS OF SIZE, BY ORGANIZING IT INTO FUNCTIONAL ZONES THAT COME AND GO AS REQUIRED.

IKEA = Invar Kamprad from Elmtaryd in the Swedish village of Agunnaryd. As founder of one of the world's most successful self-assembly furniture businesses, Kamprad is proud to admit that "profit is a wonderful word!" And according to his life philosophy, so are democracy and family values, which is why IKEA appeals to such a broad range of individuals across the economic scale.

Hoping to reach the widest public possible, from pop stars to everymen, the collection is divided into four stylistic groups: The Honest Swedish Country Life, The straightforward pragmatic Scandinavian, The Middle-class connoisseur of Modern comfort, and The Trendy Wild Young Swede. Cultural differences disappear in the clean lines and birch veneers of the flexible and funky rockers, chairs, shelves, and living ensembles.

IKEA speaks to customer/citizens who come from a variety of backgrounds but are united in values and vision. According to the IKEA marketing office, the "new generation" is composed of people who are positive, open, flexible, honest, self-critical, and willing to make mistakes. People who like their homes, have fun with them, and enjoy a life with their children. In fact, "homing" is the magic word at IKEA, whose in-store displays offer exact formulas for a cozy and secure life in a world where people's futures are uncertain.

Like a theme park, the IKEA shops have become more and more a destination where families can spend the entire day together because there's something for everyone. It's not just the furniture that makes IKEA's profit but the "satellite" accessories such as toys, cushions, duvets, tea lights, dishes, plants, and vases—impulse buys that look great (for a while), seem to be a good deal, and go well with the newly purchased sofa.

Of course trendy self-sufficiency has its price. Sure, the furniture is cheap and chic but owning it means accepting all the old stereotypes. It is no coincidence that the more functional heavy furniture, like chairs and desks, has male first names, while curtains are reserved for the ladies. Upholstered furniture (read craft/feminine) is named after Swedish villages, whiles shelves (read business/masculine) after Swedish professions.

Ever the utopian, Kamprad will stop at nothing short of a better world, which apparently just needs to be purchased and installed. With 177 branches in 31 countries, 286 million visitors a year, and a catalogue that comes out in an edition of 118 million in 23 different languages, why should Kamprad settle for less?

IKEA, sweden

1-28 NOTHOZAMILE ZAMA

zama House, ongoing
Thembani, Paarl, South Africa

TOP, BOTTOM AND OPPOSITE NOTHOZAMILE ZAMA (PICTURED, WITH ONE OF HER CHILDREN) HAS BEEN DETERMINED TO MAKE HER HOUSE, A MAKESHIFT BUILDING MADE OF ZINC SHEETING AND RECYCLED LUMBER IN THE IMPOVERISHED DISTRICT OF PAARL, SOUTH AFRICA, A COMFORTABLE AND VIBRANT HOME. WHILE RUNNING WATER AND ELECTRICITY ARE ABSENT IN THE TWO-ROOM ABODE, ZAMA'S CREATIVITY ABOUNDS. SHE HAS CREATED A HOME FROM MATERIALS HER HUSBAND BRINGS HOME FROM THE SCRAP YARD WHERE HE IS EMPLOYED. THE WALLS ARE COVERED WITH WALLPAPER MADE FROM HOME-FURNISHINGS FLYERS AND THE FLOOR IS COVERED WITH A BLUE TARPAULIN.

FOLLOWING PAGES ZAMA SITS IN FRONT OF A CABINET ASSEMBLED FROM A CRATE AND VARIOUS PIECES OF WOOD. SHE HAS CHOSEN IMAGES SPECIFIC TO EACH ROOM. SHE MADE THE WALLPAPER PASTE HERSELF OUT OF FLOUR AND WATER.

HER BABY NAPS IN THE BEDROOM, PAPERED WITH IMAGES OF BEDROOM SUITES FAR OUT OF HER PRICE RANGE.

Severe economic and social conditions force residents of the Thembani (or "have hope") district of Paarl, South Africa, to construct their homes in a makeshift fashion. Zinc sheeting and recycled lumber are the staple building materials in this deeply depressed black township outside of Cape Town. In the midst of the maze of grey and brown self-built structures, however, one house stands out: that of the family of Mrs. Nothozamile Zama. The outside, painted a royal blue with crisp white windowpanes, gives a small hint of what one can find inside.

Mrs. Zama's two-room house has no running water or electricity, but the joyfully decorated interior does not signal want or deprivation, instead, it becomes a testament to Mrs. Zama's ingenuity and resilience. Much of the material and furnishings in the house have been rescued from a scrap yard where her husband works. The walls, made of cardboard and tin from the scrap yard, have been covered with wallpaper made of magazine and newspaper advertisements from home-furnishing stores. Because she couldn't afford store-bought wallpaper glue, the paper was hung with Mrs. Zama's own homemade glue made of flour and water. In the kitchen, neatly hung advertisements of expensive kitchen dinettes and cabinets dominate, while near the sleeping area, pictures of "on sale" bedroom suites take over. Unable to afford linoleum, the ever-creative Mrs. Zama covered the dirt floor with a blue tarpaulin-like material, again rescued from the scrap yard, which she attached to the ground with evenly spaced studs made of beer caps and recycled nails. The resulting interior lives up to the intended meaning of Thembani, providing a cozy, comfortable, and one-of-a-kind home for the Zama family.

nothozamile zama , south africa

1-32 SOD HOUSES

1800s–present
Minnesota and Nebraska, USA

BELOW Affectionately referred to as "soddies," this modest yet functional sod house is located in Gothenburg, Nebraska. The distortion of the roof line is caused by the settling of the turf structure.

OPPOSITE The house is comprised of one large, open room that contains kitchen, living, and bedroom areas. The thick, bulging, sod walls are minimally "decorated" with whitewash and the ceiling is covered with a canvas tarp to catch falling dirt and bugs. The floor is plain, swept dirt. There is no electricity or running water (but there is an outhouse), and on the treeless prairie, a plentiful source of heating fuel is dried cow dung.

That scrappy frontier girl Laura Ingalls Wilder lived in one. Sod houses—or "soddies," as they are called—were common dwellings for American pioneers of the 1800s and are making a comeback with adventurous types interested in "green" architecture. They were originally created because the lack of trees, and therefore lumber, on the Great Plains made pioneers even more resourceful about suitable housing materials. But they have remained popular in certain circles because they are inexpensive to build and virtually indestructible.

Soddies are made entirely from prairie sod, which is an excellent building material because of its thick network of roots. The grass is first plowed into "bricks" of sod up to four inches thick and 18 inches wide. These bricks are then staked in layers two to three bricks deep to create walls, the same way masons use bricks to create more traditional dwellings. The bricks, with their intricate root systems still intact, are placed grass-side down, making the roots attach themselves to other bricks, locking them together. Roofs are usually made from timber, rough or planed, and covered with more sod, but can also be made from twigs, branches, bushes, and straw. Tarps are usually attached to the ceiling to catch falling particles of dirt. The interior walls of the soddie can be covered with newspaper, canvas, oilcloth, or they can be whitewashed. Hard clay or stamped dirt floors are common, making clean-up a snap: "cleaning" the floor requires watering and then smoothing it out to make a new floor. Fire isn't much of a problem because of the dirt roof and walls, and, in terms of heating and cooling, soddies tend to stay warm in the winter and cool in the summer due to their thick walls. Only a minimal amount of lumber is needed for a door and one or more windows.

Sod Houses, Minnesota and Nebraska, USA

1-34 KEN DRAIZEN
gas station house, 1998
oakland, CA, USA

ken draizen, oakland, CA, USA

When Ken Draizen, sculptor, furniture designer, and architectural metalworker, is asked to describe the renovation strategy he used for his gas-station-turned-home he chooses the term "bricolage." Another word might be "alchemy" as Draizen has a knack for turning garbage into gold. "I do a lot of scrounging," Draizen remarks. "There's so much waste in the world, I'd rather reuse things, even if it takes more time." And so with this waste-is-more aesthetic in mind, a hydraulic lift became his dining room table; the former carport is now his studio; a lightly dinged Kohler Jacuzzi tub was "rescued" from a Home Depot dumpster and installed in his bedroom; and a cozy bathroom was fashioned out of remnants of a men's washroom. The bathroom, in particular, proved challenging: Draizen left the old sink basin but soldered copper plumbing parts together to create a crook-necked faucet complete with water heater shut-off valves for taps. The toilet was separated from its tank and angled in order to fit into a corner, and, since a floor drain was already in place, the entire room became a shower enclosure. The overall aesthetic is Mad Max meets Jean Tinguely.

Before the gas station became an airy, 1,200 square-foot live/work space for Draizen and his children, the building had former lives as a neighborhood market, a bread store outlet, and a machine shop. Draizen describes his contribution to the building's history as "uncovering much of what the former owners deemed raw and unfinished and exposing and enhancing those materials with a finished aesthetic."

LEFT TO RIGHT TURNING A GAS STATION INTO A HOME ISN'T DIFFICULT WHEN YOU HAVE AN EYE FOR THE UNUSUAL AND GENTLY USED AS SCULPTOR KEN DRAIZEN DOES. DRAIZEN FOUND THE SLIGHTLY CHIPPED KOHLER JACUZZI TUB IN A DUMPSTER BEHIND HOME DEPOT AND INSTALLED IT WITH LITTLE FANFARE, EXPOSING ITS UGLY UNDERBELLY.

DRAIZEN STANDING IN HIS BATHROOM WITH HAND-FASHIONED TOWEL HOOKS MADE FROM OLD SPUD WRENCHES USED IN I-BEAM CONSTRUCTION. BECAUSE THE GAS STATION'S BATHROOM ALREADY HAD A DRAIN IN PLACE, THE ENTIRE ROOM BECAME A SHOWER.

THE ORIGINAL SINK AND TOILET WERE KEPT, BUT WERE CUSTOMIZED TO FIT DRAIZEN'S TASTE. THE TOILET WAS ANGLED TO FIT IN THE CORNER AND SEPARATED FROM ITS TANK.

1-36 LOT/EK
miller/jones studio
New York City, USA

LEFT AND FOLLOWING PAGES
IN THE HANDS OF NEW YORK–BASED ARCHITECTURE FIRM LOT/EK, GREAT REMNANTS OF INDUSTRY ARE FINDING A SECOND LIFE AS INNOVATIVE INTERIOR DESIGN. HERE A 40-FOOT LONG ALUMINUM SHIPPING CONTAINER SEPARATES A KITCHEN, BATHROOM, AND BEDROOM FROM THE WORK SPACE OPPOSITE. WHEN NOT IN USE, APPLIANCES STICK OUT FROM THE WALLS IN PLACES NORMALLY RESERVED FOR FREESTANDING FURNITURE OR DOORFRAMES.

Guided by their passion for new and innovative uses of cast-off materials, LOT/EK have made the most of industrial behemoths such as shipping containers, oil tanks, cement mixers, and the fuselage of a jet. With their *Miller/Jones Studio,* founders Ada Tolla and Giuseppe Lignano transformed a typical New York City walk-up into a flexible, functional, and practical live/work space.

Making the most of a limited ground plan, the architects divided the space lengthwise using the side of a 40-foot long aluminum shipping container. More a "containing" than a retaining wall, the hefty chunk of metal separates the private and public areas and serves as cabinet for a variety of household objects that pull out when in use. Stove, oven, sink, and refrigerator protrude from the closed wall like sculptures or cartoon furniture when not in use, but are rendered fully functional when hidden doors, cut into the metal, are lifted or opened. Rather than stripping the wall bare, LOT/EK left it exactly as they received it, signs and all. The result is unexpected but humorous juxtapositions of words and objects, such as the coincidental pairing of "Flammable Gas" with the oven.

Down the wall from the kitchen are three vertical panels that swivel open like shutters to reveal a bedroom dominated by a wardrobe of stacked filing cabinets, flat files, and gym lockers. Also embedded in one of the panels is a color TV that may be viewed from both the office and bedroom, depending on which way the panel is turned. Opposite the long wall is the open space of the rest of the studio where a long wall of windows provides illumination along with stunning views of the Manhattan skyline. During the day, this is the business end of the space and is marked by an enormous working table made of refrigerators placed on their sides. The doors of individual fridges may be propped open to form impromptu sidetables that come in handy when business hours are over and the room becomes a living area.

1-38

LOT/EK, New York city, USA

1-40 RURAL STUDIO

The Lucy House, 2001–02
Mason's Bend, Hale County, Alabama, USA

RIGHT ONE OF THE CLIENT'S PRIMARY REQUESTS WAS A PLACE WHERE SHE COULD PRAY. RURAL STUDIO COMPLIED WITH A "TWISTED" BEDROOM THAT REFERS VISUALLY TO THE REGION'S TORNADOES (AND IS ITSELF SITUATED ATOP THE HOME'S STORM SHELTER) BUT ALSO SYMBOLIZES THE SOUL OF THE DWELLING.

OPPOSITE THE HARRIS FAMILY IN THE MAIN LIVING AREA OF THEIR HOME WHICH WAS CONSTRUCTED OUT OF 72,000 INDIVIDUAL CARPET TILES DONATED TO RURAL STUDIO BY A NATIONAL CARPET MANUFACTURER. SINCE THE TILES ARE OLDER THAN SEVEN YEARS, THERE IS MINIMAL "OFF GAS" MAKING THEM PERFECT BUILDING BLOCKS.

In a bend along the Black Warrier River in Hale County, Alabama, sits an overlooked hamlet of four extended families in which makeshift shanties without plumbing facilities are home sweet home. In 1992, the community was discovered by architect and Rural Studio founder Samuel Mockbee who, with the help of architecture students from Auburn University, set out to build low-cost but noble dwellings for the residents of this ignored "pocket of poverty."

Built for clients Anderson and Lucy Harris, *The Lucy House* is emblematic of Rural Studio's commitment to honest materials, innovative thinking, and collaboration, in both design and construction, with the owners. As with other Rural Studio buildings, materials that might have ended up in a landfill—in this case 72,000 individual carpet tiles salvaged from office buildings throughout the USA—were recycled into a dwelling of architectural distinction.

Responding to the clients' desire for a storm shelter, a place to pray, and separate bedrooms for their children, the 1,200-square-foot house is composed of two parts. The main section contains three child's rooms, a bathroom, a kitchen and living area and is constructed of carpet tiles held in place by a heavy wooden ring beam. All interior walls are finished in plywood, according to student builder J. M. Tate, because the team received a donation of four pallets and "liked its durability as a material, its aesthetic appeal when placed next to the carpet, and its un-precious qualities."

A second section of the house contains an underground, poured-concrete storm cellar that doubles as a TV room and a "crumpled" sanctuary—Lucy's place to pray. Resembling a geodesic cloud, the multi-faceted form was erected in an ad-hoc way sans construction documents—the team would step back to evaluate where the next panel should be located, and each day Lucy would come to evaluate the space. When the room reached a desirable height, a window to the sky was set in and the walls were sheet rocked and painted white to emphasize its special function.

By listening to the people they build for, keeping an open mind, and experimenting with found and donated materials, Mockbee and his students have provided an "architecture of decency" to a population that time and bureaucracy forgot. *The Lucy House* is the last project directed by Mockbee before his death from cancer in December 2001.

Rural Studio, Alabama, USA

1-42

ABOVE LEFT THE UNDERGROUND, POURED-CONCRETE STORM CELLAR DOUBLES AS A MEDITATION AND TV ROOM FOR THE FAMILY. THE SPACE IS ACCESSED VIA A NARROW STAIRCASE SITUATED BETWEEN THE MAIN ENTRANCE AND THE STAIRCASE TO THE UPSTAIRS BEDROOM.

ABOVE RIGHT A VIEW INTO THE MAIN FAMILY ROOM OF THE 1,200 SQUARE FOOT HOME SHOWING THE STAIRWELL UP TO LUCY'S BEDROOM /PRAYER ROOM.

OPPOSITE AN EXAMPLE OF ONE OF THE SEPARATE BEDROOMS THE CLIENTS REQUESTED FOR EACH OF THEIR THREE CHILDREN. ALL WALLS ARE FACED WITH PLYWOOD BECAUSE OF ITS DURABILITY AND HONEST AESTHETIC.

rural studio, Alabama, USA

44 CHARLES STAGG
AV Stagg Art and Wildlife Preserve, 1981–present
Vidor, Texas, USA

charles stagg, vidor, texas, usa

LEFT CHARLES STAGG IS A PROCESS SCULPTOR WHO LIVES IN THE EAST TEXAS WOODS. PERHAPS HIS GREATEST SCULPTURE OF ALL HAS BEEN HIS COMPOUND OF BUILDINGS, THE AV STAGG AND WILDLIFE PRESERVE, NAMED AFTER HIS MOTHER. THE FIRST STRUCTURE, A HEXAGONAL, DOMED BUILDING MADE OF SHEET METAL AND CONCRETE, CAN BE SEEN IN THE BACKGROUND.

RIGHT THE AV STAGG ART AND WILDLIFE PRESERVE IS BUILT FROM GLASS BOTTLES, BEER CANS, CORRUGATED METAL, SCROUNGED BRICK, CONCRETE, CHAIN-LINK FENCE, AND OTHER SUNDRY, FOUND MATERIALS. STAGG CREATED THE COMPOUND WITH NO PRECONCEIVED PLAN, RELYING INSTEAD ON INTUITION AND WHIMSY.

In the woods outside Vidor, a small Texas town near the Louisiana border, sculptor Charles Stagg lives and works in seclusion in a home/studio he built for himself called the AV Stagg Art and Wildlife Preserve. After receiving an art education on the East coast and living there for some years, Stagg gave up the life of an urban artist, preferring to work privately in a place of his own design, far outside the range of most cultural radars.

The AV Stagg Art and Wildlife Preserve is comprised of a series of structures built from glass bottles, beer cans, corrugated metal, scrounged brick, concrete and chain-link fence, and is held together with bolted steel wire rather than nails. Described by one art historian as "what it would look like if wasps were intelligent enough to build a house," Stagg had no pre-conceived plan for the structures, preferring instead to work industriously and intuitively, letting one day's ideas feed the next. There is no running water or electricity—natural light provides the only illumination and cisterns were used to collect all the houses' water when Stagg lived there. One enters the preserve through "The Church of the Swirl," an interior the artist likens to "the inside of an elephant with it trunk spiralling up." From there, the visitor goes into the "Dog House," a space reserved for past and future canine companions, and then into a 20-foot-high circular, cone-shaped space decorated with red navigational lights embedded into the concrete walls called the "Volcano Room." Then comes the studio, the first structure the artist built, a hexagonal, domed structure made of sheet metal and concrete, that is filled to bursting with his giant DNA-shaped spiralling towers. Two "bottle rooms" (rooms whose walls are made of concrete and glass bottles)—the bedroom and sauna—complete the experience. Stagg continues to work on the preserve, adding new features as whimsy dictates. Those visitors' privileged enough to be invited to Stagg's unique preserve describe it as "a wonderful spiritual place."

FOLLOWING PAGES, LEFT TO RIGHT THE GOTHIC ARCHES SIGNAL THE ENTRANCE TO THE CHURCH OF THE SWIRL, WHICH STAGG DESCRIBES AS LOOKING LIKE "THE INSIDE OF AN ELEPHANT TRUNK SPIRALING UPWARD."

THE BEDROOM SPORTS A CURVING, 15-FOOT-LONG WALL CREATED FROM CONCRETE AND MASSIVE AMOUNTS OF GREEN, YELLOW, AND BLUE BOTTLES. THERE IS NO ELECTRICITY OR RUNNING WATER AT THE AV STAGG ART AND WILDLIFE PRESERVE: WATER IS COLLECTED IN CISTERNS AND OIL LAMPS AND WOOD FIRES PROVIDE LIGHT AND HEAT.

STAGG'S SINEWY, COLUMNAR SCULPTURES RESEMBLE STRANDS OF DNA.

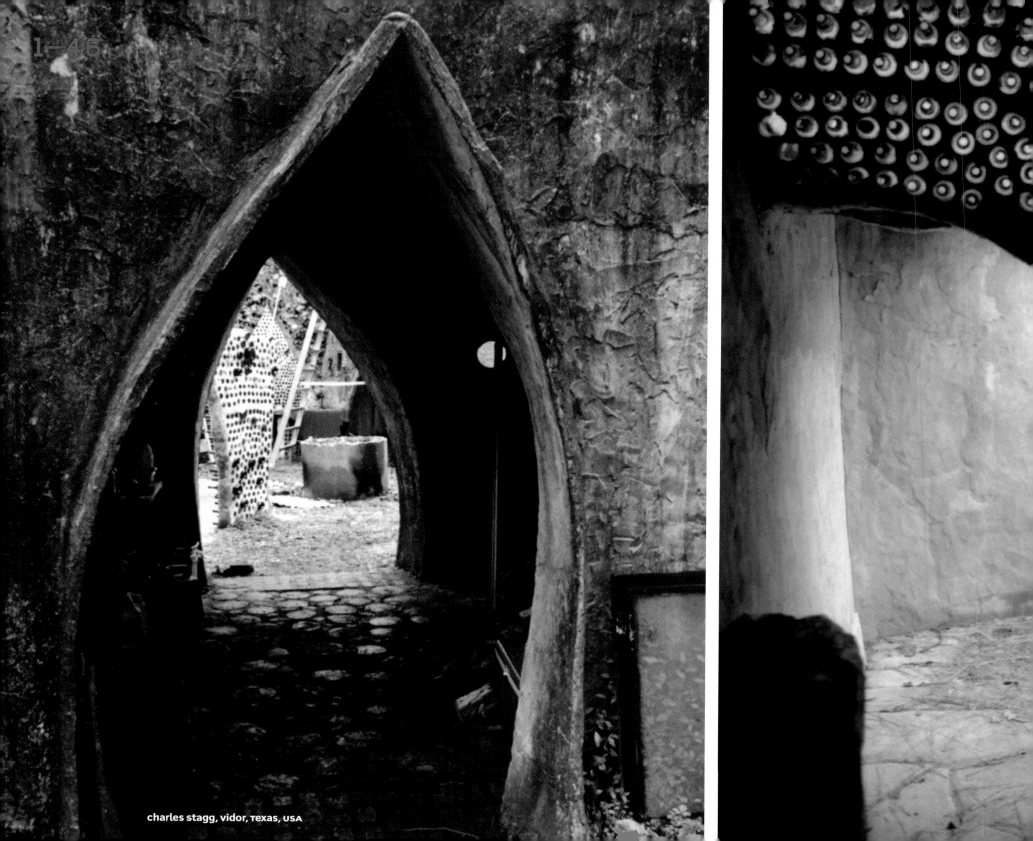

charles stagg, vidor, texas, usa

1-48 WORKHOUSE

The Workhouse, 1999 – present
Harlem, New York, USA

BOTTOM OPTING OUT OF THE NINE-TO-FIVE WORLD, "PETER" IS A TRUE FREEDOM FIGHTER WHO SPENDS HIS DAYS STRINGING BEADS ON AN IMPROMPTU PATIO OF FOUND FURNITURE OR SCOURING THE STREETS FOR NEW FINDS.

OPPOSITE THE INTERIOR LIVING SPACE OF *THE WORKHOUSE* HOLDS A BED AND DESK MADE OF CRATES AND IS DECORATED WITH FOUND OBJECTS.

An apartment, a nine-to-five day job. The trappings of middle-class life just don't appeal to Peter, who instead has chosen to reside on the street and earn his living from the deposits he collects on bottles, cans, and other recyclables. Home is a self-made shelter he calls *The Workhouse* in a vacant lot of an uptown corner of Park Avenue. It's dry, it's practical, and, best of all, it's free.

Made of 490 rescued Coca-Cola crates roofed over with an expired Dixie Chicks billboard, *The Workhouse* provides ten by twelve feet of living space with built-in shelves and "wallpaper." Not a single nail or screw has been used to hold the place together. The crates lock into place like giant Legos when stacked, and unlock with a slight twist when it's time to move along. Moving house, which Peter has already had to do once, is a largely a matter of finding the next urban slot.

The interior is accessed by a hallway on the eastern side of the building. Inside are a bed and matching shelf that have been constructed out of two large drink crates laid over four milk crates. On one wall is a shower rack holding a careful presentation of Peter's necessities and some of his "finds"—things most people take for granted but for Peter are valued *objets trouvés*. Candles and a couple of lamps, powered by electricity pinched from a lot next door, provide illumination and give the room a happy glow.

Abiding by a waste not, want not philosophy, Peter has found literally everything he feels he needs to survive: "The things people throw away...anything you need is out there."

The Workhouse, NEW YORK, USA

1-50 TSUI DESIGN AND RESEARCH INC.

Ecological House of the Future, 2002
Bao'an, Shenzen, China

RIGHT Tsui's *Ecological House of the Future* is the first ecological house ever built in China. The biomorphic shapes throughout the house belie Tsui's deep interest in evolutionary biology.

OPPOSITE The house features an indoor catchpool—constructed wetlands containing water plants that feed on bacteria. A natural sewage cleaning system, the wetlands recycle the gray and black water produced by occupants. Tsui studies the ecological relationships of living habitats and "then applies this knowledge to the design and construction of the built environment."

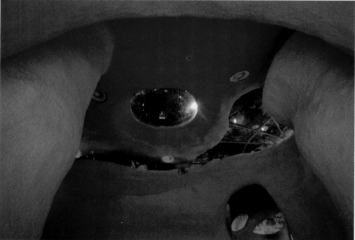

While Eugene Tsui calls the home he created in Bao'an, Shenzen, China, the *Ecological House of the Future*, its inside could easily be mistaken for a dwelling from the ancient past. The forms Dr. Tsui uses in the dwelling (as well as in his other projects) are culled from his intensive study of evolutionary biology and natural processes. Forms reminiscent of leaves, caves, shells, the curves of the inner ear abound, so while the *Ecological House* is ultra-contemporary it its green architectural concerns, its visual manifestations appears to be early Stone Age.

Located just across the border from Hong Kong, the house is the first ecological building in China and, although a private residence, it is additionally a visitation house open to the public for educational purposes. The entire structure is made of concrete, a material readily available in China that also helps to control the temperature inside. The indoor catch pool at the center of the house is a constructed wetlands that contains a selection of water plants that eat up bacteria—a natural sewage-cleaning system. The house's gray water comes up over top of building, flowing over its glass roof and into the catch pool, cooling the house while acting as a plant-feeding source. Black water from the toilet and washing machine goes directly to the catch pool where it is cleaned into gray water. The entire house is a living recycling cleaning system: water regulates the humidity in the house while at the same time cleaning the air. The house also contains hydraulically controlled, self-regulating skylights. As temperatures rise inside the house, the skylights open, letting cool breezes into the building. The floor is polished stone that contains radiant water heat, and the all the walls are topped with glass and lots of hanging planters to let in cross-lighting.

The entire structure, inside and out, is curvilinear because, as Dr. Tsui explains, "we did studies about they ways in which people move in buildings. They move in arcs and curves, not straight lines." Curvilinear structures also minimize the use of building materials while maximizing their strength. All the furnishings, except for the living room sofa, are built in, and there are no free-standing lighting fixtures (all lighting is embedded in the walls) the idea being to integrate fixtures and structures into one. The house contains three bedrooms, one large bathroom, an office space, and a kitchen that is open to the living room (which is radical for a Chinese house). But what makes the house feel so roomy is a Tsui original design: a four-foot diameter Lazy Susan closet system that extends eight feet into the air. There are four of these rotating closets, one at each "corner" of the house, which means there is not one wasted square inch in the entire house, and, enviably, never any complaints about not having enough storage space.

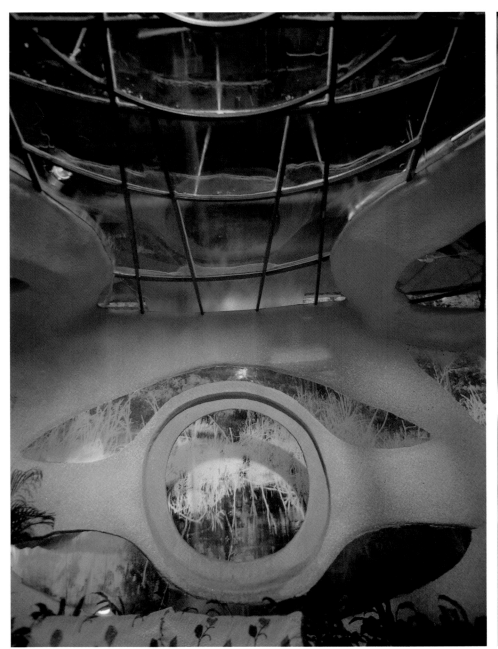

1-52

RIGHT AND OPPOSITE

THROUGHOUT THE *ECOLOGICAL HOUSE*, TSUI APPLIED HIS ARCHITECTURAL BUILDING PRINCIPLES: ECONOMY OF COST, CONSERVATION OF LABOR, APPLICATION OF NEW, SIMPLIFIED CONSTRUCTION METHODS, USE OF INNOVATIVE MATERIALS, AND ECOLOGICAL TECHNOLOGY. CONCRETE WAS THE MAJOR BUILDING MATERIAL BECAUSE IT IS READILY AVAILABLE IN CHINA AND HELPS TO CONTROL INSIDE TEMPERATURE. SELF-REGULATING SKYLIGHTS TOP THE HOME, OPENING TO CATCH COOL BREEZES, AND CLOSING TO ALLOW GRAY WATER TO CIRCULATE OVER THEM TO COOL THE HOUSE. ALL FURNITURE AND LIGHTING FIXTURES ARE BUILT-IN, AND FOUR FLOOR-TO-CEILING LAZY SUSAN CLOSET SYSTEMS USE UP THE WASTED SPACE AT THE "CORNERS OF THE HOUSE," PROVIDING AMPLE STORAGE. THE INTERIOR IS ENTIRELY CURVILINEAR BECAUSE PEOPLE "MOVE IN ARCS AND CURVES, NOT STRAIGHT LINES."

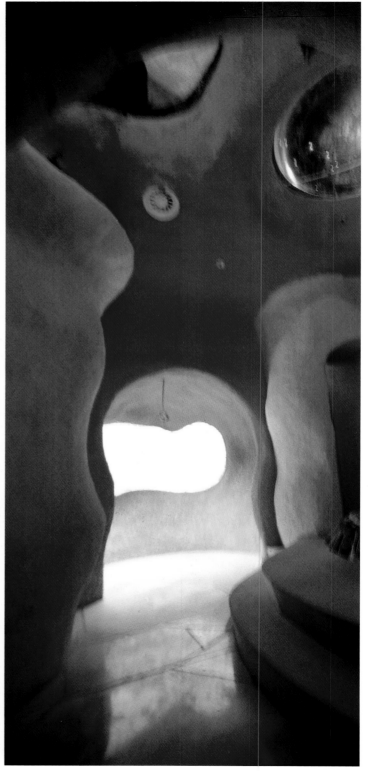

Tsui Design and research inc., china

CHAPTER 2

GO WITH THE FLOW

stasis is so over. mobility is the new black and information flow is key. sleep, play, work, eat—all these formerly separate activites now coalesce. if we are expected to multi-task, why shouldn't our homes feel the same pressure?

There was a time when a house's plan was static, immovable. Lines on an architect's drawing meant walls—the stationary kind—and each room had its separate function and attendant, appropriate furnishings. The only things that moved were the doors, and occasionally the windows. There's the kitchen, over there, the living room, one-and-a-half baths, the bedrooms. All carefully delineated, no overlap. Now, the kitchen and bathroom can change places, temporary rooms can be set up in minutes, furniture changes functions, walls morph into floors, change into ceilings, and the whole structure can be moved at will.

Today's jet-setting "global nomads" and avant-garde architects take their cues from the housing choices of nomadic cultures, those original movers and shakers who preferred huts, igloos, and tents to costly and inflexible stationary houses. Versatile, durable, and easily transportable, these dwellings were shaped by hunting and gathering, the ur-activites of civilization. In the Barak House, the architects R & Sie… looked to tent structures to create a house that not only blends into the country landscape but allows the house to expand as the needs of the family dictate, while the Swedish firm Snowcrash has developed Cloud, a compact, readily inflatable structure that provides one with a multi-purpose room within a room.

For architect Preston Scott Cohen, a stationary home can still contain aspects of mobility. Computer-generated forms provide the adventurous architect with geometric alternatives to the standard ninety-degree angles where ceiling, wall, and floor meet. Not merely a case of the "box versus the blob," Cohen's Torus House defies the architectural tradition of line as seam or boundary, dissolving familiar binaries: top/bottom, beginning/end, solid/void, inside/outside, open/closed.

NASA takes the idea of the home/office to an extreme with its *ISS TransHab prototype*. Even while traveling through several galaxies, the *TransHab* provides inhabitants with the most up-to-date technologies while simultaneously simulating the comforts of their more traditional, gravitationally challenged homes. But for those who would rather take up space than live there, there's always Berkline's Home Theater seating collection, which allows owners to travel, virtually, anywhere they wish, without ever leaving their well-padded loungers.

Designer Matali Crasset has created imaginative rooms that encourage visitors to "go with the flow" and reenergize themselves by either absorbing their recommended daily dose of light in mere minutes, or by bathing in a modern greenhouse, complete with hothouse temperatures.

The activities of artists should be watched closely as they are unusually sensitive bellwethers. During the early 1900s, the Italian Futurists proclaimed motion to be the defining force of the new century, and so it was. Now, during the early years of the 21st century, artists Urs Hartmann, Markus Wetzel, Stefan Wischnewski, and Do-Ho Suh have earmarked ephemerality, adaptability, transportability, changeability, and vernacular building strategies as hallmarks of this age. With projects such as *wildbrook, vereinsecke (club corner),* and *Apartment,* they do away with the arbitrary divisions between art, architecture, design, and life.

Taken as a whole, the inhabitants of this chapter proclaim an end to the fixed and immovable, and embrace change and mobility as the now-and-forever conditions of the contemporary domestic dwelling.

2-58 STEFAN WISCHNEWSKI

vereinsecke (club corner) and MTSV sofa, 2002
museumswinkel, erlangen, germany

as its name suggests, *vereinsecke,* by german artist stefan wischnewski, is an impromptu clubhouse decorated with the insignia, trophies, and paraphernalia of all the amateur clubs and societies of the town of erlangen, germany. inspired by the flexible and mobile architecture of trade fair booths, carnival tents, and makeshift shelters, wischnewski's *club corner* is a temporary gathering point delineated by plastic sheeting—the type that covers building materials at construction sites—and a collage of personal belongings.

with a modicum of means, the *vereinsecke* demonstrates how the most minimal and imperfect of physical boundaries can create a sense of interior space and psychological security. whether parallel lines on the ground—as in parking lots the world over—or in this case, a triangle around a scrap of carpet, private space need not be contingent on permanent walls, wooden doors, and metal locks but can just as easily occur within provisional borders that are able to move to a new location at a moment's notice.

wischnewski's *MTSV sofa* is an example of furniture that can keep up with constantly changing rooms. composed of the accessories of its conservative namesake, the männer tournee und sportverein (men's tour and sporting club), the sofa is made of soft nylon travel bags and backpacks which have been stuffed with inflated air mattresses. the bags are connected to one another with strips of velcro, allowing for a variety of combinations. infinitely more flexible than its commercial cousins—the classic corner unit, the three-person couch, and the overstuffed armchair—the *MTSV sofa* questions the static nature of the traditional living room and its attendant "family values."

at the same time, both *vereinsecke* and *MTSV sofa* reflect the paradoxical nature of today's society on the move—encompassing both wealthy jetsetters and wandering exiles—and the disturbing notion that sports fans and refugees alike are wearing the same t-shirts and using the same bags to mark their territories.

LEFT AND OPPOSITE USING PLASTIC SHEETING OF THE TYPE OFTEN FOUND ON CONSTRUCTION SITES, ARTIST STEFAN WISCHNEWSKI ILLUSTRATES THAT WALLS DON'T HAVE TO BE SOLID TO DELINEATE A COMMUNAL SPACE (SOMETHING THE JAPANESE HAVE LONG DEMONSTRATED WITH MOVING WALLS MADE OF PAPER). HERE WISCHNEWSKI MARKS TERRITORY IN THE SAME MANNER AS POPPING A TENT AND PERSONALIZES IT WITH THE LOGOS AND TROPHIES OF OVER FORTY OFFICIAL AMATEUR CLUBS IN THE TOWN OF ERLANGEN, GERMANY. DURING THE TIME THE SPACE WAS ERECTED, MEMBERS AND MUSEUM VISITORS ALIKE STOPPED BY TO HANG OUT AND WATCH VIDEOS MADE BY THE VARIOUS CLUBS.

stefan wischnewski, germany

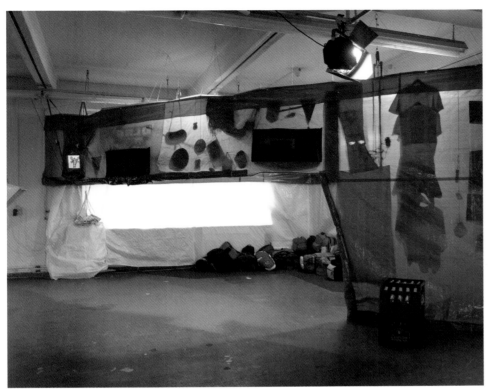

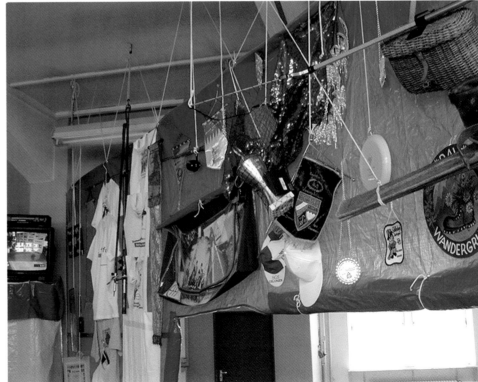

2-60

BOTTOM, RIGHT, AND OPPOSITE
MIMICKING THE CASUAL PILES OF PERSONAL BELONGINGS FOUND IN SUCH DIVERSE CONTEXTS AS GYMS, AIRPORTS, OR REFUGEE CAMPS, WISHNEWSKI HAS CREATED SCULPTURE WITH A FUNCTION. HERE TRAVEL BAGS AND BACKPACKS HELD TOGETHER WITH VELCRO STRIPS TAKE ON THE SHAPES OF SOFAS AND CORNER UNITS FOR POPULATIONS ON THE MOVE. ARTIFICIAL TURF PROVIDES A BIT OF PORTABLE LAWN.

stefan wischnewski, germany

2-62 URS HARTMANN & MARKUS WETZEL

wildbrook, 2000
zurich, switzerland

THIS PAGE ARTISTS URS HARTMANN AND MARKUS WETZEL'S PLAN FOR THE INSTALLATION OF *WILDBROOK*, EXCHANGEABLE AND PIVOTING KITCHEN AND BATHROOM UNITS DESIGNED FOR A ZURICH LOFT. THE ENTIRE STRUCTURE IS MOUNTED ON WHEELS, AND FLEXES ON PIVOTS, SO IT CAN BE MOVED TO SEVERAL "DOCKING STATIONS" WITHIN THE LOFT.

OPPOSITE, TOP AND BOTTOM THE SCULPTURAL BLOBS THAT MAKE UP *WILDBROOK* BREAK WITH THE OTHERWISE TRADITIONAL GEOMETRIES OF THE LOFT—THEIR RIOTOUS COLORS AND ORGANIC SHAPES ENERGIZE THE SPACE WITH THEIR ODDNESS.

FAR RIGHT *WILDBROOK* WAS DESIGNED "WITH AN ANTI-AESTHETIC DESIGN" IN MIND. THE RED FIBERGLASS BLOB CONTAINS THE RATHER SPACIOUS BATHROOM.

FOLLOWING PAGES DESCRIBED BY THE ARTISTS AS "TRENDY CAVES," *WILDBROOK*'S MOBILITY ALLOWS THE LOFT INHABITANTS TO ADAPT AND CHANGE THE SPACE TO ACCOMMODATE DIFFERENT SITUATIONS/NEEDS. WHILE FULLY FUNCTIONAL KITCHEN AND BATHROOM UNITS, *WILDBROOK* CAN ALSO BE SEEN AS AN ENVIRONMENTALLY SPECIFIC SCULPTURE.

when do a kitchen and bathroom function as a kitchen, a bathroom, a piece of art, a performance event, a work of interior design, and a movie set? when they are in the hands of artists urs Hartmann and Markus wetzel, of course. Normally, such rooms are asked to be just that—rooms—but when Hartmann and Wetzel were commissioned to create *wildbrook*, they did all they could to blur the lines between art, life, and interior design.

Hartmann and Wetzel built *wildbrook*, a project with exchangeable and pivoting kitchen and sanitary units for a zurich loft space, in an unused, outdoor space next to the Kunsthof over a span of four months. Kunsthof visitors were welcome to view the artists' specially built "tree house" atelier (the studio in which *wildbrook* was constructed was perched atop scaffolding) and monitor the piece's construction. At the same time, the artists were filming an adventure movie that used the construction process as part of the filmic narrative. Once construction was complete, *wildbrook* was moved to its intended loft destination and installed.

Described as "trendy caves," the sculptural blobs that make up *wildbrook* were created, as the artists explain, "with an anti-aesthetic design that focused on experimentation and innovation rather than creating an over-designed domestic environment." The more sculptural and colorful form, constructed out of translucent fiberglass, houses the bathroom, while the kitchen is contained within a timber and cardboard box. The entire structure is mounted on wheels, and flexes on pivots, so it can be moved to several "docking stations" within the loft. The mobility of the rooms allows the loft's inhabitants to adapt the space to accommodate different living situations—work, party, intimate dinner—as well as permit them to experience the space in a variety of different special groupings. While fully functional as kitchen and bathroom units, *wildbrook* also operates within the loft as an environmentally specific work of art.

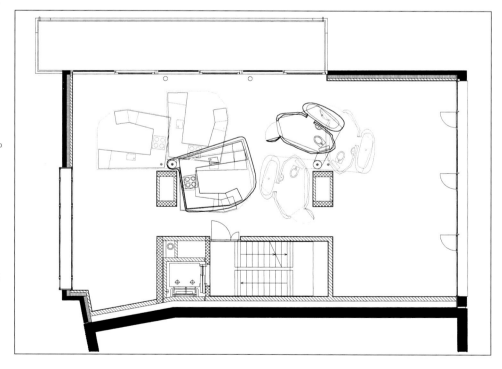

urs Hartmann & Markus Wetzel, zurich, switzerland

2-64

urs hartmann & markus wetzel, zurich, switzerland

2-66 SNOWCRASH

Founded Helsinki, Finland, now based in Stockholm
Cloud, 2002 (designed by Monica Förster)
Soundwave, 1999–2000 (designed by Teppo Asikainen)
Locations variable

Snowcrash, Helsinki, Finland now based in Stockholm

OPPOSITE CLOUD IS A PORTABLE, DREAMLIKE ROOM MADE OF NYLON AND FILLED WITH AIR. CLOUD CAN BE USED AS A ROOM WITHIN A ROOM FOR REST, MEDITATION, OR MEETINGS.

TOP LEFT CLOUD PACKS UP TIGHTLY TO BE EASILY TRANSPORTABLE AND INFLATES IN THREE MINUTES. ENTERED AND EXITED THROUGH A SELF-CLOSING SLIT DOOR, CLOUD CAN HOLD SEVERAL ADULTS COMFORTABLY.

FOLLOWING PAGES SNOWCRASH'S SOUNDWAVE SCULPTURAL WALL PANELS ABSORB, ENHANCE, OR DIFFUSE SOUND, ALLOWING THE SAVVY DESIGNER TO CHANGE THE AURAL PROPERTIES OF ANY BUILT SITUATION.

If you've always had your head in the clouds, the Finnish design firm snowcrash has the structure for you. cloud, a contemporary interpretation of Archigram's cushicle, is a portable, cumulus-shaped room that inflates in minutes and can be used as a mobile office, bedroom, meditation chamber, or for a myriad of other purposes. Monica Förster, cloud's Swedish designer, describes it as similar to a real cloud because it "goes up in the morrning and then disappears when you leave at night. It is a place where you can totally escape, but has no rules as to how you use it." Ensconced in cloud's diaphanous, airy confines, you are free to reinterpret and adapt your environment to suit your momentary needs, and when you are through using cloud, simply roll it up into its soft case and carry it to its new destination or store it away until another occasion demands it.

For those who want to improve the aural properties of their current, hard-walled surrounds, snowcrash has created a sculptural wall panel collection called soundwave. The panels, named swell, scrunch, swoop, and swell Diffuser, were designed by Finn Teppo Asikainen to enhance the acoustic properties of any interior, allowing the savvy homeowner to fine-tune each room's acoustics while maintaining a sleek, minimal aesthetic. "I like the idea of 3D wallpaper, playing with light and shadow," explains Asikainen. Descendents of Verner Panton's groovy style sans couleur, the soundwave panels were all designed with the frequency range of the human voice in mind. The swell and scrunch panels are "lightweight sound absorbers," which means they eliminate such upper frequency noises as ringing telephones and whirring computer disk drives, while the swoop panel is a "broadband absorber" and can efficiently stamp out low-frequency background noise, making individual voices more intelligible. Where swell, scrunch, and swoop act as noise absorbers, swell Diffuser operates as a sound diffuser, allowing sounds to be reflected, giving "acoustic support" to soft voices by improving speech intelligibility.

2-68

snowcrash, helsinki, finland now based in stockholm

2-70 DO-HO SUH
348 west 22nd street, apt. A, new york, NY 10011, 2001
no fixed address

RIGHT A SUIT FOR LIVING, DO-HO SUH RECREATED EVERY ROOM OF HIS NEW YORK APARTMENT IN SHIMMERING, TRANSPARENT NYLON. LIKE A FINE GARMENT, EVERY DETAIL HAS BEEN CAREFULLY PAID ATTENTION TO INCLUDING THE PLUMBING IN THE BATHROOM. WHEN NOT IN USE, THE ENTIRE SPACE CAN BE FOLDED AND PACKED FOR THE NEXT MOVE.

OPPOSITE CLOCKWISE FROM THE TOP LEFT ARE THE KITCHEN, BATHROOM, OUTSIDE CORRIDOR, AND LIVING ROOM.

when south korean-born do-ho suh relocated to the united states, his first feelings were of disconnection, dislocation, and alienation—of being a stranger in a strange place. nowadays, the forty-one-year-old artist is safe in the knowledge that wherever the path ahead may lead, he can take his home along for the ride—quite literally. a better reminder than any photograph, *348 west 22nd street, apt. A* is a life-size replica of suh's midtown apartment in new york city which he constructed en masse out of translucent nylon. from the radiator in the entrance corridor to the fireplace in the living room, kitchen appliances to bathroom plumbing, every last detail has been faithfully and painstakingly reproduced in the sheer wispy fabric.

created with help of expert korean seamstresses, the well-tailored rooms have already accompanied suh to seoul, tokyo, london, seattle, kansas city, and sydney, and may be easily folded and packed for the next move. like a favorite or well-worn article of clothing, the ensemble gives clues as to suh's personal identity while at the same time functioning as a familiar and protective second skin or a social buffer. even so, the enclosure is infinitely more public than its opaque twin, with see-through walls and fixtures that allow foreign eyes voyeuristic glimpses through and across suh's private dwelling.

In stark contrast to its real-life *doppelgänger,* suh's apartment is airy and otherworldly. walls, furnishings, and fixtures move with the slightest wind and the whole space seems to breathe like a living organism. The effect is of walking through a ghost ship in which the inhabitant's memories and spirit are still alive and present. Indeed it is suh's "desire to guard and carry around my very own intimate space" that resulted in this highly mobile but illusory dwelling.

do-ho suh, new york, usa

2-72 R & SIE

François Roche and Stéphanie Lavaux, R & Sie… / Architectes, Paris, France
Barak House, south of France, 2001

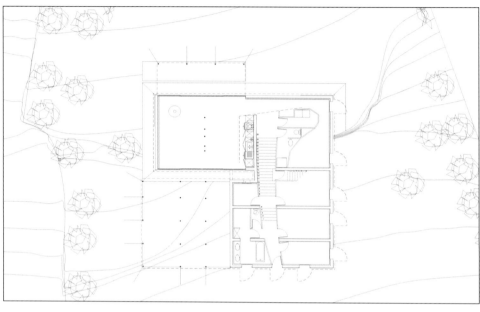

TOP RIGHT IN THE *BARAK HOUSE*, THE CONCRETE STRUCTURE CONTAINED UNDERNEATH THE TENT CONTAINS SEVEN ROOMS, WHILE THE FOOTPRINT OF THE TENT ALLOWS FOR MORE ROOMS IN THE FUTURE. THE HOUSE IS "GREEN," LITERALLY AND IN THE ENVIRONMENTAL SENSE—IT LEAVES A LIGHT FOOTPRINT ON THE GROVE OF OLIVE TREES WHERE IT IS SITED.

BOTTOM RIGHT THE REGION'S BUILDING CODES REQUIRED THAT THE *BARAK HOUSE* NOT "READ" AS A HOUSE IN THE LANDSCAPE. SO R & SIE… CREATED A "STEALTH" STRUCTURE THAT WOULD FLY UNDER THE RADAR OF AUTHORITIES.

OPPOSITE TOP WHEN THE LIGHTS ARE ON, THE GREEN TENT DRAPED OVER THE MORE PERMANENT BUILDING GIVES THE HOUSE A GHOSTLY APPEARANCE IN THE EVENING.

OPPOSITE BOTTOM THE TENT'S SHAPE, WHICH LOOKS LIKE A GIANT TOPOGRAPHICAL MAP, IS MEANT TO EMPHASIZE THE CONTOURS OF THE SURROUNDING TERRAIN. INSIDE, TRANSPARENT DRAPING IS ALSO USED TO DIVIDE SPACE WITHOUT PERMANENT WALLS. THE CONTRAST BETWEEN THE SOLIDITY OF CONCRETE AND THE DIAPHANOUS TENT MATERIAL CREATES SOME STRIKING ARCHITECTURAL COMPOSITIONS.

The program given to architects François Roche and Stéphanie Lavaux of R & Sie… for the Barak House, located in the south of France, was simple but tricky: create a house that didn't "read" as a house from a distance. In a region known for its traditional white walls and red-tiled roofs, the Barak House needed to be virtually invisible so as not to compete with the nearby historic château Sommières. Anyone wishing to build close to the château requires permission from Les Architectes des Bâtiments de France who represent the state in matters of architectural heritage and planning. So the Paris-based architecture practice of R & Sie… (pronounced in French, the letters read "hérésie," or "heresy" in English, with the three dots meaning they are open to new partners joining the firm), decided to build a "stealth" structure that would fly under the radar of architectural authorities in order to secure permission to build on protected land.

R & Sie… developed the tent image to emphasize the structure's relationship to the environment by echoing the topographical contours of the surrounding countryside. The goal, says Roche, was "to mix outside and inside." The seemingly fragile covering is made of polyurethane fabric panels clipped together with carbon fiber wires. Underneath is a concrete block structure that contains seven rooms, with enough space under the canopy to build additional rooms (also hidden from prying authorities) in the future.

The goal, says Roche, was "to mix outside and inside." And indeed R & Sie… met it. From the outside, the inside is nearly imperceptible, except at night when the lights come on and its windows cast a ghostly luminescence. And from the inside, the contrast between the solidity of concrete and the diaphanous tent material enclosing it creates some striking architectural compisitions. Winds and weather constantly change the shape of the tent, creating a flexible footprint. Inside, transparent draping, used to keep the mosquitoes at bay, also acts as a useful, permeable division between the kitchen and living areas.

In addition to blending in to the environment, the house is also "green": it has a geothermal heat source; it leaves a relatively light footprint in its grove of olive trees; and the polyurethane mesh shading the block walls helps to regulate the building's internal temperature and in stormy weather gives added protection to the construction beneath. The goal, says Roche, was "to mix outside and inside."

françois Roche and stéphanie Lavaux, Architectes, Paris, France

2-76 NASA

Lyndon B. Johnson Space Center, Houston, Texas, USA
TransHab, prototype dwelling, 1997
no fixed address

RIGHT THE ABILITY OF THE *TRANSHAB* TO HOLD AIR PRESSURE FOUR TIMES AS GREAT AS EARTH'S ATMOSPHERE WAS TESTED IN A SEVEN-STOREY VACUUM CHAMBER—ONE OF THE WORLD'S LARGEST—THAT IMITATED THE AIRLESS ENVIRONMENT OF SPACE.

OPPOSITE THE *TRANSHAB* WOULD BE CAPABLE OF HOUSING UP TO SIX ASTRONAUTS IN ITS SAFE CORE. THE INTERIOR WOULD INCLUDE SLEEPING ACCOMMODATIONS, A DINING TABLE THAT SEATS TWELVE, A KITCHEN, GYM, AND PANTRY.

For celebrities, politicians, and other VIPs, traveling within the protective confines of bulletproof vehicles is an occupational necessity. For that other group of risk takers—astronauts—safe transit could mean a voyage aboard NASA's *TransHab*, a prototype dwelling that would provide intrepid voyagers on future Mars-bound spacecraft with the hermetic security of an Earth-bound "safe room" while working in the harsh conditions of space.

Nestled within almost two-dozen layers of materials—including insulation, foam, Kevlar, and Nextel—inhabitants of the *TransHab* would be shielded from space debris traveling at more than 15,500 miles per hour (seven times faster than a bullet) and temperatures ranging from 250 degrees Fahrenheit in the sun to minus 200 degrees in the shade.

In theory, *TransHab's* durable "skin" could deflect a projectile as large as two-thirds of an inch thanks to successive layers of Nextel, a material often used as insulation under the hoods of cars. When inserted between inches-thick layers of open cell foam—similar to that of chair cushions on Earth—the two substances would cause particles to shatter and lose energy as they hit. The outer layers of the shell, in turn, would protect "bladders" of combitherm, a material borrowed from the food-packing industry, holding the capsule's air. The interior would be further strengthened by super strong Kevlar—a fabric commonly used for bulletproof clothing—embedded deep in the shell to hold the module's shape. Other features such as four-paned, four-inch windows and a water tank that doubles as a radiation storm shelter would provide protective barriers against the severe environment outside.

NASA, HOUSTON, TEXAS, USA

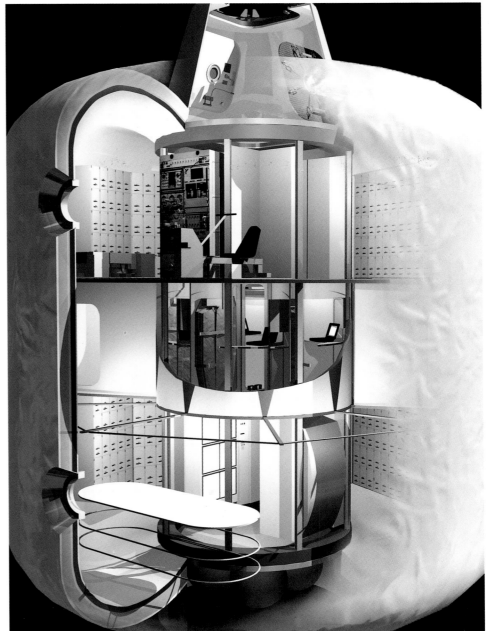

2-78

NASA, HOUSTON, TEXAS, USA

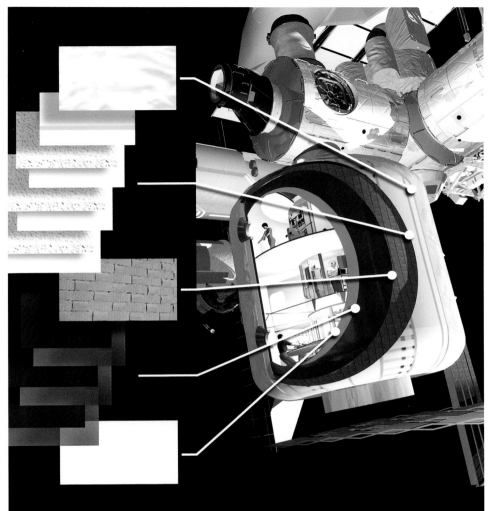

OPPOSITE A CROSS-SECTION SHOWING THE CREWS' VERTICAL SLEEPING QUARTERS AND WORK STATIONS.

ABOVE A VARIETY OF MATERIALS ACCOUNT FOR THE TRANSHAB'S "STRONG-THAN-METAL" EXTERIOR. STARTING FROM THE OUTSIDE IN ARE: AN INSULATION PANEL, ALTERNATING LAYERS OF NEXTEL CLOTH AND FOAM, A PANEL OF WOVEN KEVLAR, ALTERNATING PANELS OF KEVLAR AND COMBITHERM, AND A NOMEX INNER WALL.

2–80 PRESTON SCOTT COHEN

Boston, Mass., USA
Torus House, 2001
Old Chatham, New York, USA

RIGHT EXPRESSIONISTIC AND SCULPTURAL, THE UNDULATING INTERIOR OF THE *TORUS HOUSE* WOULD NEVER HAVE BEEN IMAGINABLE—OR AT LEAST NOT PRACTICAL TO BUILD—WITHOUT THE HELP OF ADVANCED COMPUTER SOFTWARE.

OPPOSITE PLAN SHOWING HOW THE DONUT SHAPE OF THE TORUS FLATTENS OUT INTO A SQUARE.

Walls that bend and fold, floors that rise and fall—the dynamic interior of the *Torus House* by Boston-based architect Preston Scott Cohen is more sculptural than architectural, a coalescence of smooth curves and rolling waves. Based on the principal of the torus, a donut-shaped form generated by rotating one circle along the path of a second larger circle, the house is the result of complex geometry assisted by up-to-date computer design software.

Before the digital era, such expressionistic shapes would have been unthinkable, or at least impractical to build. Today, however, architects are turning to software originally created for other purposes, such as animation, to open up new structural possibilities. Like colleagues Doug Garofalo, Frank Gehry, or Greg Lynn, Cohen relies extensively on available technologies to design his structures. At the same time, he also employs the new media to mathematically describe irregular forms, which he then merges with modern architectural principles.

This explains his co-joining of two inherently different and previously incompatible architectural forms—the "blob" and the "box"—within the *Torus House*. Built with the client's wish for two painting studios and a viewing gallery in mind, Cohen dedicated the most energetic use of interior space to art and socializing. The main living area and an easel-painting studio are given pride of place around the crooked twist of the central toroidal form. Here walls, floors, and ceilings no longer meet in tidy ninety-degree angles but fuse together in fluid and ambiguous ways—at one point the floor actually becomes a kitchen table—confusing our expectations of up and down, interior and exterior, inside and outside. By contrast, the more mundane living functions are situated in pragmatic square rooms along the home's outer edge. Regardless, it is not the exterior walls which dictate the interior design, but the preceding inner realm itself that defines the house from the inside out.

Preston Scott Cohen, Boston, Mass., USA

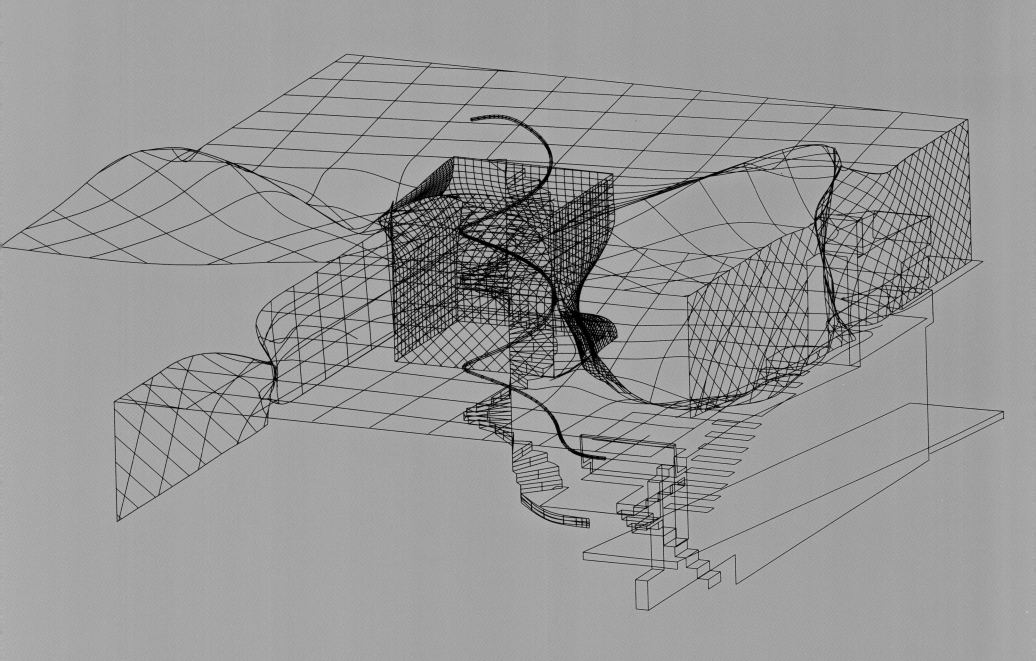

ALEKSANDRA KONOPEK

Pneo—A Foldable Habitation Module, 1999
no fixed address

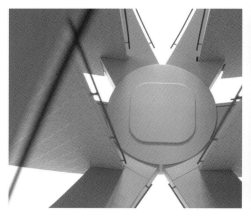 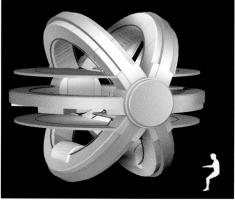 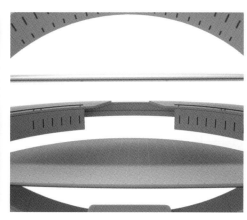

ABOVE LEFT PASSAGEWAYS AND DOCK.

ABOVE MIDDLE THE *PNEO* IN ITS FULLY INFLATED FORM.

ABOVE RIGHT VIEW OF ALL THREE LEVELS.

OPPOSITE AFLOAT IN THE LARGEST INTERIOR OF ALL, A CLUSTER OF SELF-CONTAINED *PNEOS* COMES TOGETHER IN A FORMATION THAT RESEMBLES A GRACEFUL SEA ANEMONE.

When we think of life on the final frontier, it is most likely the images generated by Hollywood than NASA that come to mind. Captain Kirk certainly manned a glamorous deck, yet for real star trekkers, the journey into space is far more banal and uncomfortable than we might imagine. As more scientists and researchers are hurled into the heavens every year, it is likely that the *Pneo—A Foldable Habitation Module* will soon become the environment of choice for astronauts seeking the comforts of home.

The way the *Pneo* module works is both simple and ingenuous. After being dropped from a space shuttle like a package into the great unknown, the module quickly unfolds and inflates into a three-ringed sphere with enough live/work space for a crew of up to six astronauts. Technical arches hold the ball in shape, allowing crewmembers to access the module's docking lock and three living areas via a passageway located in the ceiling and floor of the central level.

The central level is compact but efficient, containing the kitchen and a 360-degree multi-purpose ring-table that may be used for meeting, working, dining, and relaxing. A variety of stowed accessories, such as textile walls and mountable tables and chairs, allow the room itself to be changed multiple times, according to need or desire. Similarly, on the upper level, crew members may use textile walls to construct, separate, or change up to six private sleeping compartments. Again, the central passageway offers easy access to the shower, toilet, and exercise units nestled in the lower level.

Because the *Pneo* module is equipped with its own climate control and life support systems, it can be employed for a variety of missions and applications. Several modules could even be joined together forming spontaneous galactic flotillas. Alone or in groups, the *Pneo* is a smart and ergonomic alternative to the other heavy metal already in orbit.

Aleksandra Konopek, Wuppertal, Germany

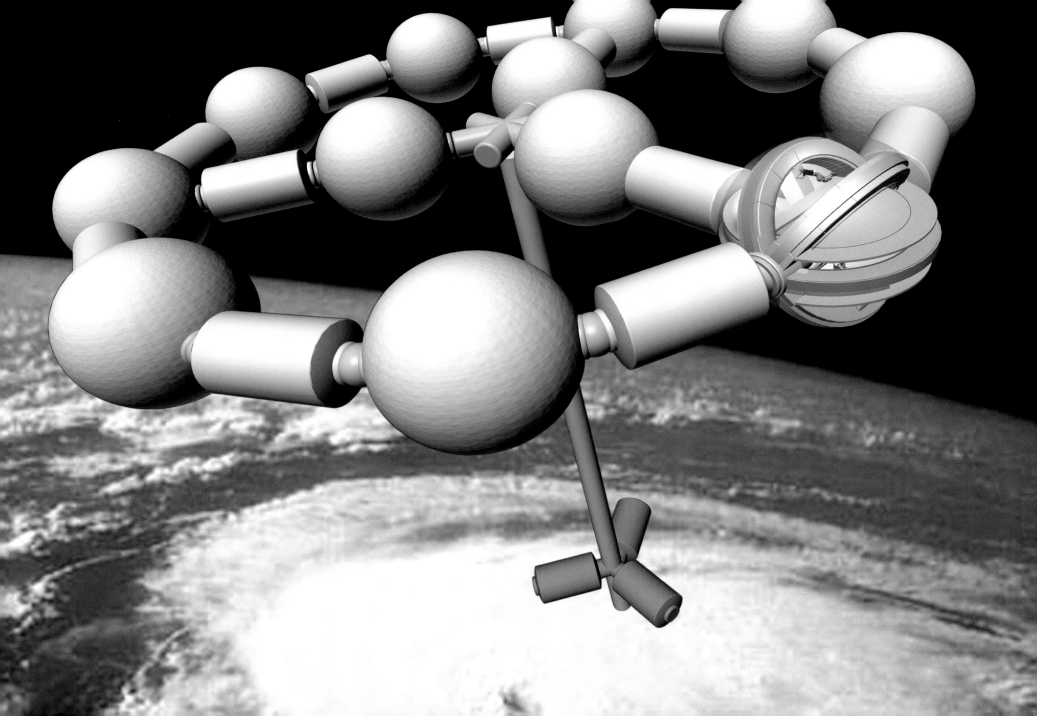

ABOVE TYPICALLY PURCHASED IN LINES OF THREE OR FOUR, THE SEATING UNITS IN BERKLINE'S *CINEMA COLLECTION* BEAR A STRANGE RESEMBLANCE TO THE MUTLI-PURPOSE, FIRST-CLASS ARMCHAIRS FOUND ON COMMERCIAL AIRLINES.

OPPOSITE THERE'S LITTLE ROOM LEFTOVER FOR OTHER FURNISHING (OR FUNCTIONS) ONCE *HOME THEATER* FURNITURE TAKES CENTER STAGE. POISED TO TAKE OVER THE SOFA AND COFFEE TABLE AS THE FURNITURE OF CHOICE, *HOME THEATER* FURNITURE COULD MAKE THE LIVING ROOM OBSOLETE—OR AT LEAST TURN IT INTO A PLACE WHERE CONSUMING FOOD AND IMAGES IS THE ONLY ACTIVITY.

The paradox of today's mobile society is that it is not necessary to actually travel or move one's body to experience cultures near and far. The virtual realities generated by Hollywood combined with the endless stream of information blurted out by the TV have contributed to the false impression that seeing is "being there."

True, television has always been a member of the traditional family unit, but with the ascent of DVDs, flat-screen TV, and surround sound, the family room in which it holds pride of place is quickly turning from a gathering place into a media mogul's screening room. If the marketing of *Home Theater* seating by companies such as Berkline, Inc. and La-Z-Boy are anything to judge by, the days of communing with the group have changed to days of reclining passively with the group to absorb prepackaged messages and cathode rays of cinematic proportion. For people who want the experience of staying put to resemble what they would experience by going out, as the Berkline web site explains, "there's nothing better than being 'front row and center' in your own home."

Marketed in free-standing units of three or four modules outfitted with cupholders, eating trays, and a variety of reclining options (*PowerRecline*, *WallAway*, and *TouchMotion*, to name a few), Berkline's *cinema collection* seating can be configured by the user into short or long rows, and even placed on raised platforms to enhance the sensation of being in a theater. A defining feature of every model is padding—in every possible area that could come into contact with the body—and some models are even available with *Tempur-Pedic* cushioning that conforms to the human shape like a mattress.

Differentiated from a bed by context only, the seating offers its users the pampered and extreme comfort normally reserved for hospital patients and the ultra rich, that is, the opportunity to lounge for hours on end with only brief excursions to the restroom or kitchen. Ironically called "motion" furniture in the industry (because it reclines, not because it sits in front of motion pictures), *Home Theater* seating is coming to visit—and staying indefinitely—in a home near you.

BERKLINE INC., MORRISTOWN, TENN., USA

2-86 MATALI CRASSET

Paris, France
Energizer Room and *Phytolab*, 2002
Designed for Dornbracht, Germany

ABOVE RIGHT The designer sitting inside *Energizer*, part of a series of innovative domestic spaces commissioned by Dornbracht, a German faucet manufacturer. She describes the space as one to "spend four or five minutes [in] to regain a dose of energy, let's even say a dose of optimism."

BOTTOM RIGHT AND OPPOSITE Views of *Phytolab*, a room that is devoted to bathing. Plants are placed in a modernist grid, filling the room with oxygen and scents. To Crasset, bathing is a combination of "hygiene and aesthetics."

FOLLOWING PAGES Visitors to *Energizer* are asked to climb into the hanging plastic suit and remain, suspended, for a few minutes, in order to receive their daily dose of light.

When approached by Dornbracht—German faucet-makers and patrons of bathing-relating art—to imaginatively reinterpret the bathroom, French designer Matali Crasset imagined the space not as one exclusively used for cleansing, but rather as a space that can also offer refreshment and rejuvenation. A former designer in Philippe Starck's über-hip design atelier and a rising star in French design circles, Crasset created rooms that awaken all the senses as well as being imaginative, whimsical, sensual, and futuristic.

In *Energizer*, Crasset interprets light as a source of well-being. Crasset invites light-lovers, or those suffering from seasonal affective disorder, to absorb a day's dose of light and energy in a few minutes by literally bathing themselves in light. The visitor to *Energizer* dons a protective space suit and floats within the air-filled pod, with arms outstretched, while artificial sunlight stimulates his or her mind and spirit. Four generators, placed at the corners, wash the room in a warm, yellow light. Switches on the generators allow the visitor to choose their required light levels.

In *Phytolab*, Crasset creates a room dedicated entirely to bathing. Surrounded by a minimalist grid of luscious green plants in a greenhouse environment, the bather is encouraged to think of the therapeutic aspects of plants while cleansing: plants as medicinal extracts, soothing tea, oxygen producers. The hothouse-like humidity promotes the growth of the plants, and their growth, in turn, promotes the overall air quality of the space.

Matali Crasset, Paris, France

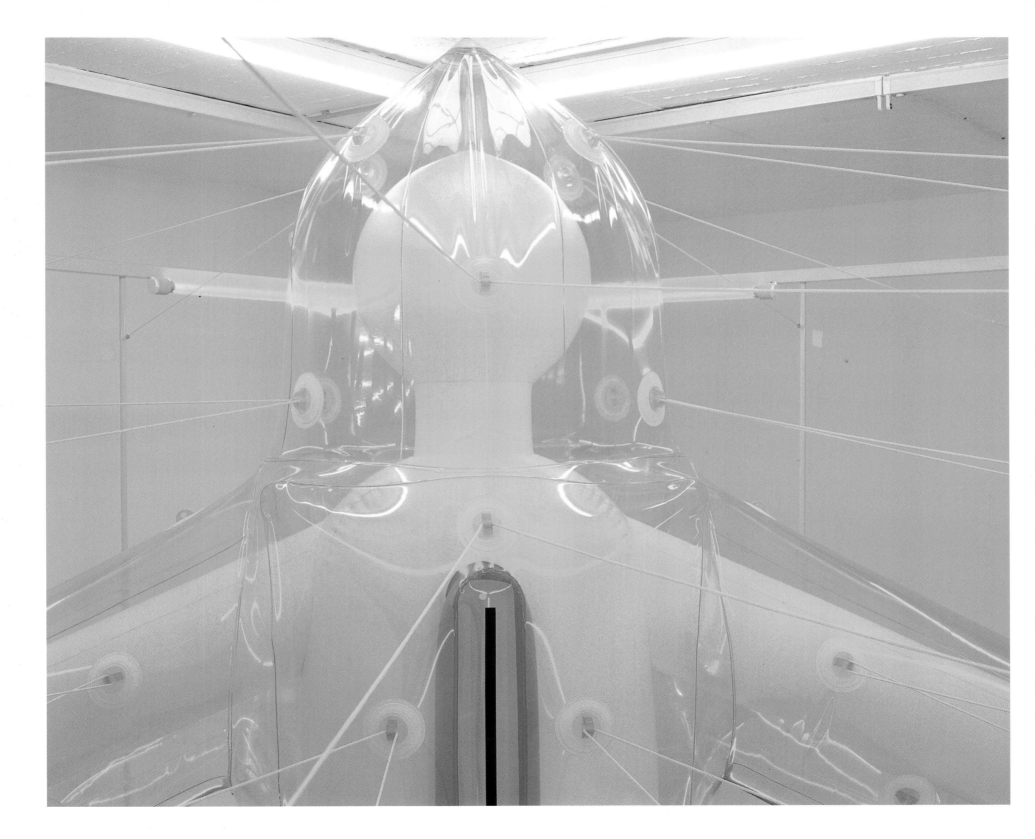

CHAPTER 3

MOVING PICTURES

pictures give us access into every conceivable nook and cranny of life and today's designers, architects, artists, and other space-makers are taking note. If an image can be cut and pasted, enlarged, reduced, compressed, cropped, adjusted, manipulated, or copied, why not space as well?

This chapter explores how the camera and the computer have opened up new ways of perceiving space. There's the glossy "head shot" prevalent in so many architecture magazines. The bird's-eye view. The close up. The detail. The long view. The almost pornographic angles from above or below. Rather than thinking about space as a void on a blueprint, today's designers, architects, and artists are conceiving of it as a collage of elements, an optical illusion, a pattern, a stage set, a moment pulled from flux, or even straight documentation.

    Today's interiors are not just bland boxes with clear borders but pictorial, even cinematic, experiences. In several instances in this chapter, rooms are better described as disappearing acts, where ceilings, floors, and furniture meld together into seamless wholes. As in a painting or a photograph, this is achieved through an all-over color scheme or pattern that causes shapes to bend and blend optically. Artists have typically been the forerunners in picturing space and the ones highlighted herein are no exception. Inspired by early modernist abstraction, English artist Maurice Agis has reproduced its two-dimensional forms and colors in large-scale inflatable environments, while Berlin-based artist Hans Hemmert democratizes his living spaces by encasing them and their contents in fine membranes of yellow latex. Another Berlin artist, Marion Eichmann, has recently completed an exercise in total immersion by wrapping a room and its contents in a knitted black-and-white wave pattern reminiscent of Op Art paintings. Eichmann's room dissolves into a continuous background, which is also true of French designer Ennemlaghi's all-white apartment and Erve Architecture's house of glass, in which interior spaces lose their borders, flatten out, and become optical planes.

    Another tendency is that our interiors are starting to take on the pictorial conventions of dioramas, theater backdrops, or movie sets—in other words, they are being imagined and composed in the all-front-and-no-back manner of a scenic painting or tableau. For Korean-born Seoungwon Won, the computer has enabled her to scan, import, and combine images from wildly different contexts into unified fantasy scapes where nature and architecture overlap and merge on a printed surface. The scenes are so real they might pass for found photos of existing utopias. A change of scenery, on the other hand, is exactly what New York set designer Reno Dakota had in mind when he transformed his apartment into an aging 1930s mansion by replicating the outward signs of decay. With expert faux finishing, Dakota successfully disguised his East Village walk-up into an image from another era. The U.S. Army's *Force Provider,* however, is very much an image of this day and age. A containerized city that may be deployed anywhere in the world, the sprawling assemblage of tents looks as much like a Hollywood back lot or set for a war movie as it does an actual theater of operations.

      The thing about pictures is that they seem to tell the whole story all at one time. They provide an all-in-one, encompassing view. This is especially true of computer-generated images that show space from all sides at once. Exploring the ways in which architecture and digital media interact, London designers Softroom have come up with a completely "rendered" space that for the moment exists only as a picture, while Florian Wallnöfer's pocket penthouse is an interior so complex that it can only be experienced in its totality as a series of photos.

3-94 HANS HEMMERT

Berlin, Germany
Samstag Nachmittag zuhause in Neukölln (Saturday Afternoon at Home in Neukölln), 1995
Unterwegs (On the Way), 1996
Im Atelier (In the Studio), 1997

Hans Hemmert, Berlin, Germany

OPPOSITE *SAMSTAG NACHMITTAG ZUHAUSE IN NEUKOELLN* (SATURDAY AFTERNOON AT HOME IN NEUKOELLN), 1995. THE BOY IN THE PLASTIC BUBBLE: BERLIN ARTIST HANS HEMMERT LOUNGES INSIDE HIS HOME WHICH HE HAS COMPLETELY ENCASED IN A THIN MEMBRANE OF YELLOW LATEX.

ABOVE RIGHT *IM ATELIER* (IN THE STUDIO), 1996. THE ALL-ENCOMPASSING YELLOW COLOR TRANSFORMS THE ARTIST'S STUDIO INTO A CAVERNOUS ENVIRONMENT WHERE EVERYDAY OBJECTS TAKE ON THE CHARACTERISTICS OF STALAGMITES.

FOLLOWING PAGES *UNTERWEGS* (ON THE WAY), 1996.

For a series of photo-works in light-boxes made between 1995 and 1997, Berlin-based artist Hans Hemmert performed and documented an experiment in total expansion. Using a fine membrane of opaque yellow latex, which he inflated with air, Hemmert compressed the interiors of his home, car, and studio and photographed the results. The effects are simultaneously playful and humorous, menacing and alienating, depending on the observer's point of view.

In related works, Hemmert has inserted himself into giant yellow latex "balls" and posed for photographs while performing tasks such as sitting on a scooter, holding a baby, or climbing a ladder. Completely encapsulated and isolated in the yellow wombs, Hemmert relied on a sharpened sense of touch to make contact with the external world. Even so, viewers on the outside saw only alien bubbles attempting to forge relationships with their surroundings to awkward and comical ends.

With the photos of room installations, however, Hemmert invites the viewer inside the "bubble" to experience a strange atmosphere situated within a familiar one. For these projects, membranes were inflated within preexisting spaces—sort of like a balloon being blown up inside a hand. As the membranes expanded, they conformed, of necessity, to every available nook and cranny, stretching taught and locking the objects underneath into a vacuum-packed topography. Walls, ceilings, and floors were unified into seamless environments, and everyday objects were transformed into mysterious mounds, lumps, or edges suggestive of strange and libidinous uses.

Simultaneously warm, all-over wonderlands that any child would love and claustrophobic, sensory deprivation tanks fit for an asylum, Hemmert's rubbery landscapes are provocative studies in tension that question the distinctions between inside and outside, good and evil, safety and danger, escape and entrapment, life and death.

Hans Hemmert, Berlin, Germany

3–98 MARION EICHMANN

16.324.800 Maschen (16,324,800 stitches), 2002
no fixed address

RIGHT AND OPPOSITE "THE IDEA OF THIS WORK CAME FROM TAKING PHOTOS IN BERLIN, PHOTOS OF EVERYDAY SITUATIONS IN TYPICAL BERLIN AREAS AND OF TYPICAL BERLIN PEOPLE. I MADE COLLAGES OUT OF THESE PHOTOS WITH VERY SMALL INTERRUPTIONS SO THAT…THE QUESTION WAS ALWAYS OF SCALE AND DIMENSION.…I TRIED TO LEAD BACK THESE QUESTIONS OF SCALE AND DIMENSION INTO A ROOM INSTALLATION [WHERE THEY WOULD BECOME] MORE CONFUSED AND EXPERIMENTAL."

One way of gaining your bearings in a foreign space is to scan the room for details as if looking at a photograph. Being subsumed by the larger picture, however, is what happens to viewers upon entering the knitted installation by Berlin artist Marion Eichmann. Filling roughly 12 square meters and covering ceiling, walls, and floor, *16,234,800 stitches* is simultaneously a spatial collage and disorienting environment spawned by Eichmann's penchant for making photographs and her curiosity as to how a two-dimensional image can become an interior space.

Developed for her master's degree in textile and surface design at the Academy of Art in Berlin (and later exhibited at the Vitra Design Museum in Berlin and Weil-am-Rhein), the room took four-and-a-half months to complete with the artist working 12 hours a day, seven days a week. Covered from top to bottom in undulating black-and-white "waves," the room's borders dissolve and flatten out like a picture. "The idea was to find a pattern that was neutral and confusing at the same time," Eichmann explains, so that the viewer can no longer find any fixed point within the room itself.

To heighten the effect, Eichmann also camouflaged a spoon, a cup and saucer, an armchair, a table and chair, a vase of flowers, and a pair of shoes with the same wavy knit, forcing them into an equal relationship with the room's architecture and allowing them to be discovered in much the same "pop up" way as images in a Magic Eye print or an Op-Art painting. By covering every object in the room with the same pattern, "an abstract situation of space develops" in which objects lose their utilitarian functions, melt into the background, and are difficult to detect by the viewer even though he or she is in the familiar conventional setting of a "room."

In Eichmann's self-spun world, interiority resides close to the surface and is best experienced optically. Like her 1960s predecessor, Japanese artist Yayoi Kusama, who disguised rooms and objects with large polka dots, Eichmann disrupts our preconceived notions of space by tampering with its appearance.

Marion Eichmann, Berlin, Germany

3-100 FLORIAN WALLNÖFER, POOL ARCHITEKTUR

vienna, austria
t.o.'s space (penthouse apartment), 1999
(owner: johannes rudnicki)

LEFT FLORIAN WALLNÖFER CREATED A DIMINUTIVE PENTHOUSE APARTMENT—USING THE FOOTPRINT OF THE WATER TANK THAT USED TO BE IN ITS PLACE—THAT, NONETHELESS, FEELS ULTRA-SPACIOUS. THE KEY IS INGENIOUS SPACE USAGE, WHICH ALLOWS THE SINGLE ROOM TO COMFORTABLY CONTAIN A CLOSET, DINING TABLE, KITCHEN, BED, AND LIVING ROOM.

OPPOSITE VARIOUS PERMUTATIONS OF THE PENTHOUSE. TWO LIGHTWEIGHT WOOD CONTAINERS ARE KEY: THEY STORE AWAY THE BED, KITCHEN TABLE, AND CLOSET WHEN NOT IN USE. THE BED AND TABLE NEST TOGETHER IN ONE CONTAINER, WHILE THE CLOSET HIDES IN THE OTHER. WHEN NEEDED, EACH CAN BE PULLED OUT SEPARATELY, FREEING THE AREA OF UNNECESSARY FURNISHINGS. TO FURTHER SAVE SPACE, THE REFRIGERATOR IS HUNG FROM THE CEILING AND THE SHOWER/BATHROOM SINK ARE NESTLED NEARBY. THE ROOM CONTAINING THE TOILET IS THE ONLY OTHER ROOM IN THE HOUSE.

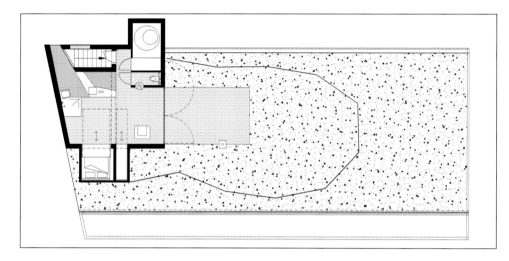

when johannes rudnicki, whose family owns a five-story building he converted from a dairy factory into office spaces for artists, architects, and designers in the 1990, wanted to add on a penthouse apartment for himself, he was dismayed to discover that the roof wouldn't support any more weight than that of the existing water tank. so he turned to pool architektur, residents in his building, for help. florian wallnöfer, one of the partners of pool (others include christoph lammerhuber, axel linemayr, and evelyn rudnicki), took inspiration from the water tank, using its footprint as that of the apartment.

although diminutive by traditional penthouse standards at a mere 193 square feet, wallnöfer created a single room that comfortably and ingeniously contains a closet, dining table, kitchen, and bed—one room that effectively serves as bedroom, bathroom, kitchen, and living room. key to understanding how one room can function as many, are the two lightweight wood containers that rudnicki and his friends created that are tucked into the south wall. these store away the bed, kitchen table, and closet when not in use. the bed and table nest together in one container, while the closet hides in the other. when needed, each can be pulled out separately, thereby freeing the space of unnecessary furnishings.

parallel to the "bedroom/dining room" is the "kitchen/ shower." along a steeply angled wall, a cooktop, sink, ceiling-mounted refrigerator, and showerhead are installed. the shower basin is a stainless-steel tray sunk into the floor. only the toilet is given a separate room in a small closet, but even it becomes part of the main room through a rotating drum in the wall adjoining the "living room" that lets one watch television while using the facilities. while such tiny spaces tend to feel claustrophobic, wallnöfer has created a penthouse that has an expansive feel due to its smart design and a floor-to-ceiling glass wall that opens onto a rooftop terrace, providing a breathtaking view over the city.

florian wallnöfer, vienna, austria

OPPOSITE AND BELOW THE PENTHOUSE FEELS EXTRA SPACIOUS BECAUSE OF ONE WALL THAT IS MADE OF GLASS. THE "WALL" OPENS UP TO A LARGE ROOFTOP PATIO AND PROVIDES THE OWNER WITH AN EXPANSIVE VIEW OF THE ROOFTOPS OF VIENNA.

3-104 NATICK LABS (U.S. ARMY)

Massachusetts, USA
Force Provider, 1991–present

RIGHT THE U.S. ARMY'S FORCE PROVIDER IS A ONE-SIZE-FITS-ALL BASE-IN-A-BOX. THE 10-ACRE BASE CAMP HOUSES 550 SOLDIERS IN COMFORT.

OPPOSITE EACH MOVABLE CITY CONSISTS OF 100 OR SO CONTAINERS, TAKES ABOUT TEN DAYS TO ASSEMBLE, AND NEEDS 50 SUPPORT STAFF TO SET IT UP AND RUN IT. FORCE PROVIDER COMES IN AN OFF-THE-SHELF PACKAGE WITH INSTRUCTIONS INCLUDED, AND CAN BE ASSEMBLED ANYWHERE BY ANYBODY, THEREBY ELIMINATING THE NEED FOR QUARTERMASTER BATTALIONS. TODAY'S SOLDIERS ARE DEPLOYED WITH ALL THE COMFORTS OF HOME: SATELLITE TELEVISIONS, CHAPELS, "ADVANCED" SHOWERS, LATRINES, LAUNDRY ROOMS, AND COMPLETE KITCHENS AS WELL AS CLIMATE-CONTROLLED TENTS.

Forget about army bases of yore, with their rough and tumble aesthetic, leaky tents, and sad latrines. The American soldier of today expects all the comforts of home when he or she is shipped off to war. Enter *Force Provider*, the newest housing effort by the U.S. Army—a one-size-fits-all base-in-a-box. The Army's Natick Labs in Massachusetts began designing the kits in 1991 after G.I.s in operation Desert Storm complained about living conditions. The Army borrowed the idea of a "base-in-a-box" from the Air Force, which has kits called *Harvest Eagle* and *Harvest Falcon* for its soldiers. Designed to make military operations more efficient by providing "containerized, rapidly deployable cities" for quick delivery to theaters of action, the 10-acre base camp, which takes about ten days to assemble, needs 50 support staff to set it up and run it. Because *Force Provider* comes in an off-the-shelf package with instructions included, it can be assembled anywhere by anybody, thereby eliminating the need for quartermaster battalions. Each movable city consists of 100 or so containers, and houses 550 soldiers in comfort, with satellite televisions, chapels, "advanced" showers, latrines, laundry rooms, and complete kitchens as well as climate-controlled tents. And, each $5-million module is designed to work in weather ranging from 15 below zero or 120 above. "[The soldieries] have got just about everything you've got at home, except a wife," says William Oliver, a supervisor at the Defense Depot in Albany, Georgia where some of these kits are stored. The Army has deployed 36 *Force Provider* modules to various locations around the world to be set up at a moment's notice.

3-106 SOFTROOM

London, England
maison canif, concept for wallpaper* magazine, 1997

softroom, london, england

LEFT AND RIGHT THE TERM "ALL-IN-ONE" TAKES ON NEW MEANING IN THE *MAISON CANIF*, A VIRTUAL INTERIOR MODELED ON A SWISS ARMY KNIFE. THE DIFFERENT ZONES—LIVING ROOM, STUDY, DEN, BEDROOM, KITCHEN, AND BATHROOM—ARE DESIGNED TO BE OPEN OUT FROM FOUR CORNER PIVOTS WHEN DESIRED OR REQUIRED.

The London-based, multi-disciplinary design studio softroom combines "the practice of built architecture with the exploration of digital space." This sometimes means working in the form of "concept" projects such as *masion canif*, which allow the group to appropriate the possibilities of the digital world to potential real-life living environments.

Maison canif is a virtual prototype for a self-contained, all-in-one interior (one unit, one control panel) that, in theory, could be inserted into any preexisting room or building, here in a redundant office space. Based on the design of a Swiss Army knife, the living areas are contained in various "blades" which may be pulled out from the titanium host singly, in groups, or out all at once, depending on need or the duration of the task. Rather than relying on perimeter walls to provide a sense of interiority, *maison canif* has no set floor plan or ground rules.

For the sake of practicality, everyday functions such as socializing, eating, washing, and sleeping are grouped loosely around four corner pivots. Built-in features such as remote control make it easy to move, say, the sofa from a position close to the unit to a location further a field and with a better view. Flexible storage units, which also double as luggage, may be combined with fixed furniture to produce an intimate "den." Likewise, a variety of cooking/dining arrangements may be established by simply swinging out combinations of prep surfaces, appliances, and seats. Even the bath may be extracted independently or together with a smoked-glass screen to form a shower enclosure.

Technologies such as flip-up TV and electronic fireplace give the space the signs of a more conventional home, while a diagnostic touch-screen displays the configuration of furniture and keeps the individual parts networked.

3-108

RIGHT AND FAR RIGHT WITH THE STORAGE UNITS PUSHED OUTWARD, A COMBINATION OF FIXED AND SWUNG FURNITURE CREATES A DEN. A DIAGNOSTIC TOUCH-SCREEN BELOW THE HEARTH (FAR RIGHT) DISPLAYS THE CURRENT PATTERN OF FURNITURE AND THE TEMPERATURE. LIKE THE POSITION OF A DRIVER'S SEAT IN A LUXURY AUTOMOBILE, IT MAY BE PROGRAMMED FOR LATER RECALL.

softroom, London, England

3-110 KRUUNENBERG VAN DER ERVE

The Netherlands
Laminata House, 2001
Leerdam, The Netherlands

RIGHT EXTERIOR VIEWS OF *LAMINATA*, THE FIRST HOUSE TO USE GLASS AS A BUILDING MATERIAL. OVER 13,000 SHEETS OF GLASS WERE LAMINATED TOGETHER FOR STRUCTURAL STRENGTH. THE HOUSE GLOWS FROM WITHIN LIKE AN AQUARIUM.

OPPOSITE WALKING THROUGH THE HOUSE IS LIKE WALKING UNDERWATER, THE THICK SHEETS OF GLASS LANGUOROUSLY REFLECTING LIGHT. THE WALLS VARY IN THICKNESS CAUSING THE COLORS TO RANGE FROM ALMOST CLEAR, THROUGH VARYING SHADES OF TRANSLUCENT GREEN TO A DARK, NEARLY BLACK, GREEN. IRONICALLY, REAL WINDOWS WERE KEPT TO A MINIMUM TO PRESERVE STRUCTURAL INTEGRITY.

FOLLOWING PAGE, LEFT PART OF THE *LAMINATA HOUSE* IS A "REGULAR" HOUSING STRUCTURE: ITS CONCRETE FOUNDATION CONTAINS A GARAGE, WORKSHOP, STUDY, AND SMALL KITCHEN.

FOLLOWING PAGE, RIGHT IN THE BEDROOM, A SINGLE, TRANSPARENT SHEET OF GLASS STRETCHES ACROSS THE FULL WIDTH OF THE BUILDING'S SHORT END. THE CONTRAST BETWEEN THE THICK CORNERS AND THE THIN, LARGE EXPANSE OF GLAZING IS STARTLING.

The ethereal, immaterial qualities of glass have intrigued architects for centuries. Architects in the Gothic period strove for the largest possible expanses of stained glass in their cathedrals; Joseph Paxton's nineteenth-century wonder, the Crystal Palace, astounded world Fair goers with its vast glass ceiling; while in the twentieth century, Mies van der Rohe and Philip Johnson set about creating their own versions of the perfect minimalist glass house. But when Gerard Kruunenberg, Paul van der Erve, and Oliver Thill of Erve Architecture turned their attention to glass, it was not with the intention of setting large panes of glass within structural metal frames. Rather, they wanted to use glass as a building material in and of itself. The result is the *Laminata House*, the first genuine glass residence.

One might expect an all-glass house to be the ultimate transparent shelter, but *Laminata* is the exact opposite. Using over 13,000 sheets of glass that have been laminated together for structural strength (hence the name), the building is an opaque, shimmering sea green box. The project took four years of research alone, and, in the end, used over 13,000 panes of glass, each catalogued and cut via computer, then carefully glued together into laminated "blocks." Each block is ten millimeters thick and measures 3.2 meters by 6 meters. The panes were glued together to create two 20-meter-long rectangular masses, which were placed 6 meters apart. Three rooms were created in the gap—the entrance, the patio and the living room, separated by four single panes. The main bedroom and guest bedroom are in the larger box while the smaller volume contains a corridor. Doorways were cut out of the walls and windows were kept to a minimum, with just a few long narrow slots cut out of the thick walls. The house sits on a concrete foundation box that contains a garage, workshop, a study, and small kitchen.

Walking through the house is like moving through frozen water. Especially stunning is the angular corridor that runs the length of the building. Because the walls vary in thickness, the colors range from almost clear, through varying shades of translucent green to a dark, nearly black, green. In these shifts of color and gradations of translucency, the *Laminata House* turns traditional perceptions of glass inside out: glass becomes a concealer, rather than a revealer, of domesticity.

Kruunenberg van der erve, the Netherlands

3-114 ENNEMLAGHI, LTD.

Paris, France
white apartment, 2001
Paris, France

Ennemlaghi, Ltd., Paris, France

LEFT THE WHOLE APARTMENT IS COVERED IN WHITE RESIN—EVEN THE OVERHANGING KITCHEN ISLAND. UNDER THE SPECIALLY DESIGNED RESIN COUNTER HIDES A SINK, OVEN, REFRIGERATOR, AND WASHER/DRYER (THE COUNTERTOP SLIDES BACK WHEN NEEDED). A SHALLOW BASIN OF WATER HOVERS IN THE COUNTERTOP.

RIGHT ENNEMLAGHI WANTED TO CREATE A PURE, CLEAN SPACE, "LIKE LIVING INSIDE A PORCELAIN CUP." ALL THE FUNCTIONAL ASPECTS OF THE APARTMENT ARE HIDDEN AWAY, AND ONLY THE OWNER KNOWS HOW TO ACCESS THEM ALL. ENNEMLAGHI STRESSES THAT THE COLOR CHOSEN WAS NOT IMPORTANT—IT COULD HAVE BEEN YELLOW, BLUE, ANYTHING. THE VITAL ASPECTS OF THE DESIGN ARE THE TOTALITY OF CONCEPT AND MATERIALS, AND THE PRIVACY THE SPACE PROVIDES.

It's become a Hollywood stereotype to portray Heaven as an all-white space, like an infinitely large photography studio backdrop—no discernable edges, no shadows, great ambient lighting. Similarly, modernists such as Mies van der Rohe and Richard Meier have used white in their architecture to suggest purity of thought and lines, a steadfast devotion to the strictures of minimalism, and an ascetic's proclivity towards the intellectualism of white versus the emotionalism of color. But outside of a Hollywood sound stage or an architectural model, not many would choose to live everyday lives in pristine white confines. An all-white space demands of its inhabitants a certain rigor, that's why when Ennemlaghi, Ltd., a Paris-based "virtual design group," presented their proposal for an all-white apartment to a fashion designer-client, they specified that their proposal be either wholly accepted or rejected—there would be no middle ground.

What their client got for his agreement was a U-shaped space in Paris' Bastille neighborhood that has been completely clad in white resin. All traditional room divisions have been minimized and all things utilitarian such as the kitchen sink have been provided with plastic tops so they can be covered when not in use. Ennemlaghi even designed the living room furniture—clear plastic-covered sofa and chair(s) that were manufactured according to artisanal methods used since the 18th-century—that leave their metal springs exposed. Ennemlaghi believed that the owner's own possessions would be enough to give the space a personalized touch. The effect of the interior is zen retreat meets avant-garde art gallery meets chic boutique, or as the apartment's inhabitant describes the experience of living there, "on n'a plus de repères ici (here you lose your bearings)."

3-116   RENO DAKOTA
Apartment, 1983–present
NEW YORK CITY, USA

RIGHT EVEN THE REAL WOODEN BLINDS IN THE KITCHEN RECEIVED THE "AGE RAVAGED" LOOK. DAKOTA DISTRESSED VARIOUS SLATS BY HAND.

FAR RIGHT A DETAIL OF THE MADLY PATTERNED ROCOCO KITCHEN FLOOR.

OPPOSITE IN RENO DAKOTA'S NEW YORK TENEMENT APARTMENT, THERE ARE NO LIVING ROOM WINDOWS, SO HE CREATED THREE, REPLETE WITH FAKE BLINDS AND CURTAINS, OUT OF WALLPAPER AND BLIND MOLDINGS. THE SET DESIGNER AND FILMMAKER DESCRIBES HIS PLACE AS MIMICKING AN AGING 1930S MANOR IN A STYLE HE HAS DEEMED "DECAY DÉCOR DAKOTA."

To Reno Dakota, all the world's a potential stage design, including his own home. The New York set designer and filmmaker, who also styles props and designs for commercial photography, has lived in his small East Village walk-up since 1983. Unsuspecting visitors should be forgiven, though, if they mistake his tenement apartment for an aging 1930s manor—such is the completeness of Dakota's illusion. His decorating style, which he has deemed "Decay Décor Dakota," is a complex study in faux finishing. In the pattern-on-pattern living room, which he calls the "organic chamber" and describes as "what a living room should look like," there are no living-room windows, only artfully applied layers of wallpaper that have been fitted into gilded, blind moldings. Floral patterned wallpapers vibrate against geometric patterned wallpapers, as demure, tasselled furniture provides the only rest for the eyes. Even the baseboards have been carefully thought about—they are artfully chipped to reveal years of overpainting by various landlords. In the kitchen, each slat of the wooden blinds has been carefully sawed to look age-ravaged and the floor has been covered with an intricate mosaic of flooring fragments from the 30s, 40s, and 50s that he scavenged from the garbage. Dakota's illusion extends to his bedroom, too. Also known as the "Geometric Room," Dakota used wallpaper to create a frieze of rectangles on the ceiling, matching swatches of one paper with the same pattern in a brighter colorway, so that the first pattern seems sun-faded.

Dakota cites conceptual artist David Ireland and design diva Diana Vreeland as the major influences in his apartment design, as well as his love of the look of decay: "[Decay] reminds us that humankind will inevitably lose everything to the forces of nature. And, I say, the sooner the bunnies and dinosaurs take over, the better."

RENO DAKOTA, NEW YORK CITY, USA

3-118

RENO DAKOTA, NEW YORK CITY, USA

LEFT THE DOORWAY CONNECTING THE LIVING ROOM TO THE KITCHEN DISPLAYS SEVERAL CONTRASTING WALLPAPER PATTERNS. HIS OLD GE MONITOR TOP REFRIGERATOR IS IN TIP-TOP CONDITION AND RESTS ON A FLOOR CREATED FROM A COMPLEX INTARSIA OF LINOLEUM FRAGMENTS FROM THE 30S, 40S, AND 50S THAT HE "RESCUED" FROM THE TRASH.

RIGHT DAKOTA GILDED THE FLOORBOARDS AND THEN, ARTFULLY, TOOK A HAMMER TO THEM, CHIPPING AWAY YEARS OF OVER PAINTING BY LANDLORDS. HE LOVES THE LOOK OF DECAY, MUSING, "[DECAY] REMINDS US THAT HUMANKIND WILL INEVITABLY LOSE EVERYTHING TO THE FORCES OF NATURE. AND, I SAY, THE SOONER THE BUNNIES AND DINOSAURS TAKE OVER, THE BETTER."

3-120 SEOUNGWON WON

Düsseldorf, Germany
Wunschzimmer – Jörg, 2001
Wunschzimmer – Michalis, 2002
Wunschzimmer – Zushun, 2001
Wunschzimmer – Fabian und Eliane, 2001

RIGHT *WUNSCHZIMMER – JÖRG*, 2001. KOREAN ARTIST SEOUNGWON WON USES THE CAMERA AND COMPUTER TO "PAINT" UTOPIAN WORLDS IN WHICH MAN AND NATURE COEXIST IN HARMONY. IN HER VIRTUAL DIORAMAS, ROCKS BECOME PIECE OF FURNITURE AND COLUMNS RISE UP LIKE TREES.

FOLLOWING PAGES
WUNSCHZIMMER – MICHALIS, 2002 (LEFT); *WUNSCHZIMMER – ZUSHUN*, 2001 (TOP RIGHT); *WUNSCHZIMMER – FABIAN UND ELIANE*, 2001 (BOTTOM RIGHT).

When discussing interior design, images of contemporary zoos rarely come to mind. But this is a comparison that can't be ignored when viewing the photomontages of South Korean artist Seoungwon Won. Entitled *wunschzimmer*, German for "wish rooms," Won's photos are surreal, computer-assisted collages of scenes from nature, architectural elements, furniture, human figures, and all sorts of creatures great and small. Cut, pasted, and merged into a new digital reality, the scenes depict a bucolic life in which animals, people, and the trappings of modern life happily coexist.

How Won composes these complex images is where the crossover to today's zoos becomes apparent. In some cases, the photos could easily be mistaken for actual documentation of the weird and unexpected ways that nature and culture collide in the park. In all of Won's images, human beings are presented as if on stage. That is, centered in the picture and framed on three sides by architectural elements and natural wonders. A well-known old master painting device, the triangular composition, focuses all attention and action on the foreground.

In much the same way, zoo architecture is geared towards presentation, with the "domestic" areas of the animals arranged into tableaux so that the animals' daily lives become high drama worthy of public viewing. Real rock formations merge freely with man-made constructions, water flows over concrete streambeds, and animals roam between wooded "center stages" to concrete bunker "backstages." All the while, the necessities of zoo maintenance crop up in strange and disillusioning ways: electric fences snake through the shrubbery, cables crawl along rock walls, metal doors open out of hills, windows peek from stone outcrops.

In a similar manner, Won's worlds depict the strange and wonderful hybrid that occurs when man communes with nature while hanging onto the comforts of modernization. Beds, bookshelves, and chairs rest in water as though stones, walls with curtained windows crop up in green fields. Here a man takes a dip in his own living room, there a woman showers in a personal waterfall. Won's are indeed wonderful habitats, ideal interiors in utopian worlds that unlike dioramas in zoos—perhaps fortunately for us—can for the moment only exist in the virtual space made possible by the camera and computer.

Seoungwon Won, Düsseldorf, Germany

seoungwon won, düsseldorf, germany

3-124 MAURICE AGIS

dreamspace, 2000–01
London, England

English artist Maurice Agis has been creating what he has termed "three-directional" artworks since the 1960s. Agis, inspired by the constructivist and De Stijl schools of abstraction, decided to take their often two-dimensional, planar experiments with form and color into the area of sculptural space. Initially collaborating with fellow artist Peter Jones, the two sought to create sculptural spaces that viewers could move through and interact with, escaping the chaos and fragmentation of the senses that city life can foster. Agis was frustrated, however, that the construction of these room-sized sculptures relied on an already built environment, like a gallery or museum, to contain them, and so when artists went their separate ways, Agis began researching inflatables, as were other artists and groups of that period: Archigram, Graham Stevens, and the Eventstructure Research Group.

What Agis discovered with these individual artistic and architectural groups was an appreciation for the mobility, pliability, affordability, and changeability of pneumatic structures. Now Agis' sculptures could escape the confines of the gallery and fully exist in the public realm. In his recent walk-in sculpture, *dreamspace*, Agis creates a psychedelic, forest-like thicket of chambers by connecting numerous multi-colored plastic cells. Visitors are invited to don colored capes to enhance the over-all effect and encouraged to move freely through the maze of tunnels. Completely cut off from the outside world, visitors are enveloped in reds, blues, greens, and yellows while specially composed music swirls around them, enhancing the totality of the experience. The air currents flowing through the skin of the sculpture, holding it up, and the ebb and flow of natural light that illuminates the space, gives dreamspace a living, breathing effect. The more one immerses him or herself in totality of the experience, the more blurred the boundaries between inside and outside, dream and reality, spectator and artwork become, and the closer Agis comes to his desire to create a universally apprehended art experience.

LEFT AN AERIAL VIEW OF ONE OF AGIS' PNEUMATIC SCULPTURAL ENVIRONMENTS. AGIS PREFERS INFLATABLE PVC STRUCTURES BECAUSE THEY ARE MOBILE, INEXPENSIVE, AND THEY HAVE THE ABILITY TO TRANSFORM ANY LOCATION IN WHICH THEY ARE ERECTED.

RIGHT WITH THEIR PSYCHEDELIC COLORS AND THE INVITATION FOR VISITORS TO FULLY IMMERSE THEMSELVES IN THE EXPERIENCE, MAURICE AGIS' *DREAMSPACES* ARE SURELY SOMETHING AUSTIN POWERS WOULD CALL "GROOVY, BABY!" IN MORE RECENT *DREAMSCAPES*, AGIS HAS SPECIALLY COMPOSED MUSIC PLAYING AND ASKS VISITORS TO DON COLORED CAPES TO ENHANCE THE TOTALITY OF THE EXPERIENCE. PEOPLE ARE ENCOURAGED TO WANDER AIMLESSLY THROUGH THE FOREST-LIKE MAZE OF TUNNELS.

Maurice Agis, London, England

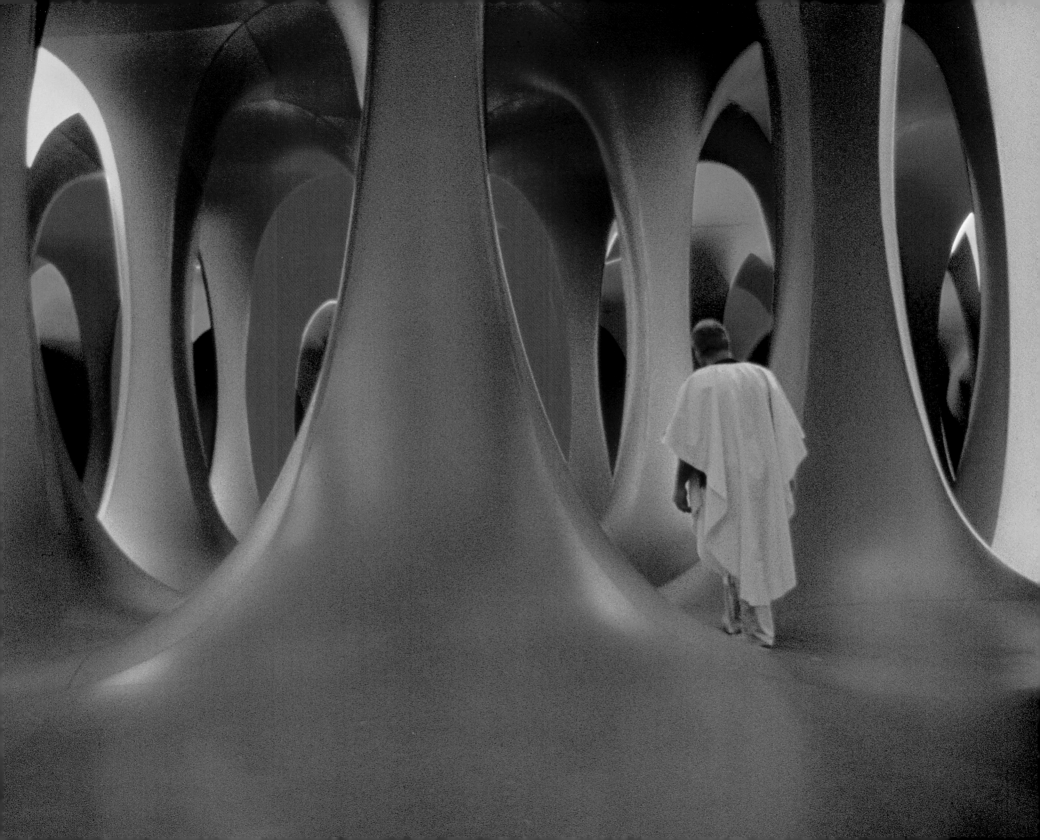

CHAPTER 4

CONTENTS UNDER PRESSURE

once, our private dwelling spaces differed significantly from penal institutions or psychiatric wards, but none of these categories is mutually exclusive anymore. Today's domestic interiors bear many of the marks of incarceration and surveillance, while jail cells and mental wards are becoming more domestic, even user-friendly.

Whether imposed on us by others or ourselves, "discipline" is a term that unites many of the projects in this chapter. A marked change in the way control is imposed on those behind bars or in solitary confinement is evident in the sleek and minimal designs for prison facilities by the DLR Group (USA) and the upbeat color schemes and accessories for an isolation room in a Swiss hospital by Francisco Torres. Both environments look almost homey and function like mini apartments, complete with the same compact and versatile furniture marketed to inner-city dwellers. Designer Ammar Eloueini and his students at the Chicago-based Archeworks have taken this approach a step further by recommending the rehabilitation of prisoners in communal outdoor settings.

On the other end of the spectrum is the phenomenon that inmates aren't the only ones under twenty-four-hour surveillance. Private citizens are increasingly scrutinized by hidden eyes, be they security cameras in the cities where we live or, more invasively, self-installed monitors in the spaces we call home. American security guru Al Corbi has made a fortune catering to our fear of the bad guys—real or imagined—and in today's climate of global unrest, political instability, and sporadic terrorism his panic rooms, Kevlar mattresses, in-house monitors, and built-in weapons systems are an easy sell. Paranoia and xenophobia are on the rise and nowhere more evident than in Corbi's own mansion outfitted with nine million dollars worth of defense systems against missile, rocket, and biological weapon attacks.

Another area of crossover occurs where our interiors merge with virtual reality, as demonstrated by the wildly successful "reality" television show, Big Brother. The show's producer, Endemol, has found no end to the number of people ready and willing to bunk with perfect strangers in a sealed container under the watchful eye of television and web cameras. Contestants gladly forfeit personal liberties and access to the outside world in exchange for one hundred days in an isolated "nowhere" carefully observed by each other and millions of foreign viewers.

The media has also placed enormous pressure on us to live the good life, to be forever young. Beautiful homes and beautiful bodies fill our screens and magazines, although Dutch designer Lucas Maassen has thrown the equation out of balance in his study, *conflict of proportion.* Alarmed by how fluid the borders between childhood and adulthood are becoming (parents and children often dress alike nowadays), he illustrates the unnatural disjunction that occurs when a four-year-old girl occupies the coded space of an adult—in this instance a kitchen—that has been scaled to the proportions of her own body.

Seclusion is also a theme that crops up in this chapter in interiors used as escape hatches. Both Atelier Van Lieshout's *sleep/study skull* and FAT's *camo house* accommodate "the sleeper," in both its senses. For AVL, there's not much more to life than sleeping and working, working and sleeping, so why should an interior contain anything more (or less) than a bed and desk? On the other hand, London-based FAT touch on the more contemporary and sinister use of the word in their camouflaged family room that could easily be the meeting place of a local militia.

Last but not least, there is an overwhelming tendency to compress space—and our bodies with it—beyond the bounds of normalcy, whether to test our physical limits or to simply adhere to building codes. Artists Absalon and Gregor Schneider have subjected themselves to rigorous dwellings of their own making—a sterile white pod and a house within a house, respectively—in an effort to achieve self-awareness and perfection. Likewise, American artist Chris Burden and Japanese architects Masaki Endoh & Masahiro Ikeda exploit the borders of tiny vertical spaces on crowded city lots that are so compact inhabitants must adhere their bodies and routines to the tight, irregular spaces rather than vice versa.

4-130 FAT (FASHION, ARCHITECTURE, TASTE)

London, England
Camo House, 1998

RIGHT AND OPPOSITE PATTERN IS A CARRIER OF PERSONAL TASTE AND HAS AS MUCH TO DO WITH THE WAY WE PERCEIVE OUR INTERIORS AS BUILT WALLS. ACCORDING TO LONDON-BASED ARCHITECTS, FAT, TASTE PLAYS A CRUCIAL ROLE IN THE CONSTRUCTION OF SPACE AND IS "FAR MORE RADICAL THAN THE SPATIAL GYMNASTICS FAVORED BY THE MAINSTREAM ARCHITECTURAL AVANT-GUARD" AND IS "FAR MORE ENGAGING TO A WIDER CULTURE BEYOND THE ARCHITECTURAL ACADEMY."

with so many images of war and conflict streaming through our television screens these days, why shouldn't the drapes and sofa follow suit? For London-based collective FAT, the coding of space is a fertile area of social and architectural investigation. operating under the slogan "taste not space," the group argues that taste is precisely the point where "architecture engages with issues of class and values and hence is the moment when architecture becomes politicized."

Looking like a cross between a *Good Housekeeping* advert and a propaganda poster for the armed forces, *Camo House* is one in a series of digitally enabled interiors that asks why practitioners of "high" architecture trivialize the role of taste. By changing the decorating scheme from Dusty Rose to Desert Storm, FAT raises all sorts of questions about the role of fashion, the foremost being: can pattern ever be just pretty?

As if to answer, a heavenly, peaceful light pours in through the room's virtual window, illuminating a middle-class setting swathed in super-sized camouflage. The image is as heartwarming as it is an uncanny precursor to this decade's global unrest and an ironic commentary on the commingling of religious iconography and weaponry in homes both East and West. (Jesus and Mohammed alike watch over gun cabinets and mounted swords.) simultaneously bold and funky, oppressive and aggressive, it's hard to say whether the pattern is menacing or simply the latest "fad" in home decorating. After all, love of family and nation are mainstream values, a fact demonstrated by the periodic recurrence of the military "look" on designer catwalks.

As FAT maintains, taste has everything to do with defining space and "the recoding of a very everyday home with an alien pattern serves to illustrate the potential of surface decoration in opposition to 'truth to materials'... favored by Modern hegemony."

FAT, London, England

4-132 ATELIER VAN LIESHOUT

The Netherlands
sleep/study skull, 1996
Tampa skull, 1998

RIGHT ATELIER VAN LIESHOUT PRODUCES OBJECTS THAT CROSS THE BOUNDARIES BETWEEN ART, LIFE, AND DESIGN. HERE, AN INSTALLATION VIEW OF A *SLEEP/STUDY SKULL*, ONE OF THE COLLECTIVE'S SCULPTURES THAT ARE MEANT TO FACILITATE A "SELF-SUFFICIENT AND INDEPENDENT LIFESTYLE."

OPPOSITE THE CLEAN LINES AND STRAIGHTFORWARD USE OF MATERIALS SHOWS AVL'S DEBT TO MODERNISM, WHILE THE AUSTERE INTERIORS ALSO REFERENCE MONKS' CELLS AND SENSORY DEPRIVATION CHAMBERS.

When Rotterdam-based Josep van Lieshout started Atelier van Lieshout (AVL) in 1995, it was with the intention of creating standardized, made-to-order furniture. Since that time, the multidisciplinary and collaborative AVL, under van Lieshout's direction, has shifted its attention to creating works of art that can also be used to facilitate "self-sufficient and independent lifestyles" such as mobile homes, office units, sensory deprivation pods, and other quirky furnishings. These works, which carry visual traces of van Lieshout's interest in Modernism in their clean, minimal forms, nevertheless possess quirky, handcrafted qualities that also suggest a survivalist mentality.

Like fellow designers/artists Dré Wapanaar and Andrea Zittel, who also create customized architectural sculptures and are concerned with issues of self-sufficiency and isolation, van Lieshout's enterprising sculptures fall somewhere between utopian fantasy and utilitarian sculpture, blurring the once-rigid boundaries between art, architecture, and design. His mutant mobile unit, *Tampa skull*, with its jagged, blue exoskeleton, provides a containerization of living spaces. The compact, modular 7 x 7 x 25-foot space, made of PVC and wood, contains a kitchen, living area, den, toilet and, at the far end, a bedroom. This is a space for an individual to retreat and rest, alone, in peace and comfort. In the six-foot-long brown *sleep/study skull*, which contains only a bed and desk, the Atelier has pared down possible activities to a mere two, signaling monkish devotion and asceticism and begging the question, "what else is there to do but work and sleep and sleep and work? The titles derive from the shapes of earlier versions of the modules, which were like lumpy skulls, in addition to being a humorous comment on the possibility of truly "living in one's head."

Atelier van Lieshout, The Netherlands

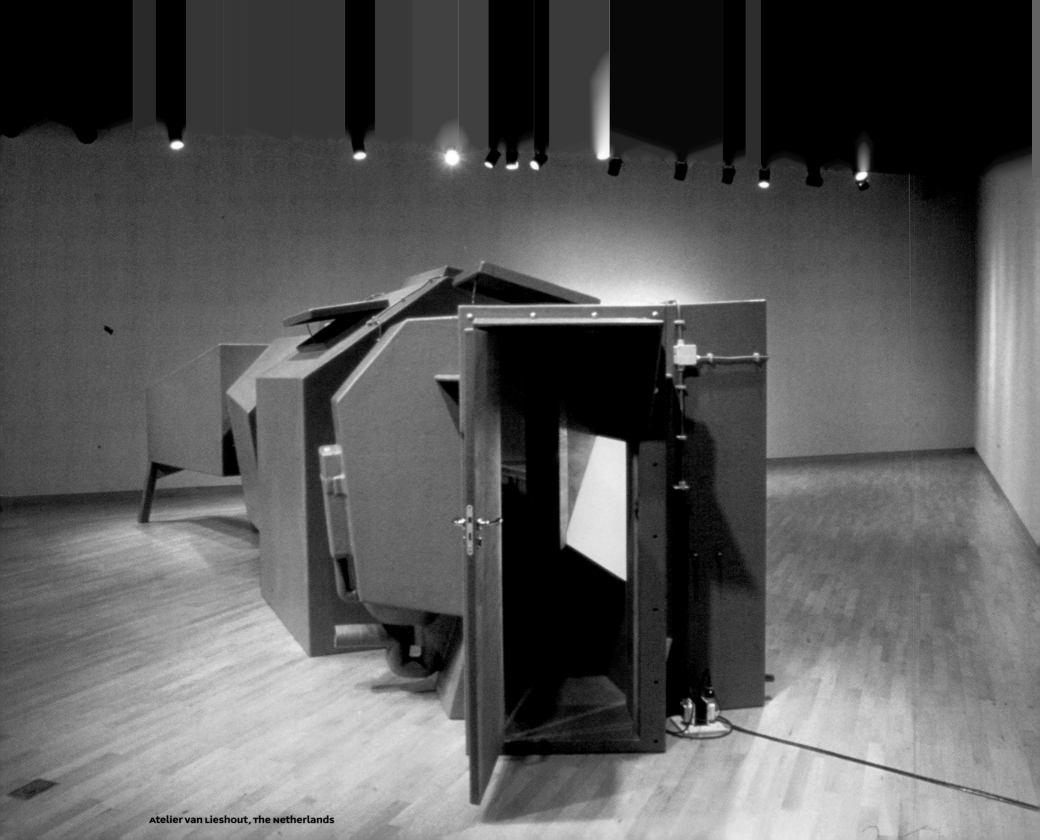

Atelier van Lieshout, The Netherlands

LEFT THE COMPACT SPACE DOESN'T ALLOW FOR MANY ACTIVITIES OUTSIDE OF THE WORK'S CONCISE TITLE.

RIGHT THE *SLEEP/STUDY SKULLS* ARE MEANT AS ESCAPE PODS FOR INDIVIDUALS SEEKING REST, DESIRING MEDITATION, OR JUST SOME QUIET TIME ALONE.

4-136 DLR GROUP
Jail Cell, Auckland Central Remand Prison, Auckland, New Zealand

ARCHEWORKS
Entropias, prototypes for prison cells, 2000–01, Chicago, IL, USA

RIGHT THE AUCKLAND CENTRAL REMAND PRISON IN NEW ZEALAND, CREATED BY THE DLR GROUP, IS AN EXAMPLE OF THE MOST UP-TO-DATE THINKING ABOUT INTERIOR DESIGN STRATEGIES FOR CORRECTION FACILITIES. INDIVIDUAL CELLS, OUTFITTED WITH MODERNIST-LOOKING, WALL-MOUNTED, STAINLESS STEEL FURNITURE, LOOK AS IF THEY COME FROM IKEA. GENERAL CIRCULATION AREAS ARE BRIGHTLY COLORED, SLEEK, AND MINIMALLY FURNISHED, IN KEEPING WITH TODAY'S DESIRE FOR A MORE "RELAXED," DORMITORY-LIKE PRISON ENVIRONMENT.

RIGHT MORE THEORETICALLY, THE STUDENTS OF ARCHEWORKS, AN "ALTERNATIVE DESIGN SCHOOL" IN CHICAGO, QUESTION TRADITIONAL IDEAS OF CRIMINAL CONTAINMENT IN THEIR *ENTROPIAS* PROJECT.

ACROSS TOP AND MIDDLE TOP LEFT DRAWING LARGELY FROM MICHEL FOUCAULT'S TEXT, *DISCIPLINE AND PUNISH*, THE STUDENTS RE-IMAGINE PRISON LIFE AS NOT HIDDEN AWAY FROM THE PUBLIC BUT RETURNING TO THE PUBLIC AS SPECTACLE. IN THE *HAMSTER CAGE*, PRISONERS WOULD WALK SIDE-BY-SIDE WITH CITY DWELLERS IN TRANSPARENT TUNNELS, WHILE THE *GLASS PRISON* WOULD EXPOSE ALL THE FORMERLY OBFUSCATED ACTIVITIES OF PRISON.

MIDDLE TOP RIGHT AND LOWER MIDDLE THE *IRIS* CONCEPT IS A FORCED ENVIRONMENTALIST FACILITY. ITS GOAL IS TO REHABILITATE INMATES ALONG WITH THE ENVIRONMENT. SOME ARE "EXILED," HERE SEEN SUNK INTO LAKE MICHIGAN, WHILE OTHERS ARE "INTEGRATED" INTO THE URBAN STRUCTURE.

BOTTOM OTHER *ENTROPIAS* IDEAS FOCUSED ON THE PANOPTICON AND ITS ABILITY TO PROVIDE CONSTANT SURVEILLANCE AND THE STRUCTURAL PROPERTIES OF MOLECULES.

Although prisons have existed since Roman times, it's not until the eighteenth century that prisons become houses of correction and discipline, designed with the intention of punishing and reforming prisoners through isolation, deprivation, and confinement, rather than just being places of torture. Punishment, to some extent, has over the years become more psychological than corporeal, and prison architecture and interior design reflect this desire for psychological control. In the 1800s, the circular plan of Jeremy Bentham's *Panopticon* (where the cells are lined up along the circumference of the space with an ever watchful guard tower in the center) and the radial plan (where halls of cells radiated out from a central core assuring the solitary confinement of prisoners) were popular design models. But even today, where a more "relaxed," dormitory-style design becomes the favored model, the issue of how best to house criminals remains a hotly contested issue.

Since 1966, the DLR Group of architects has grappled with the housing needs of the criminal justice market. One of their most recent projects, the Auckland Central Remand Prison in Auckland, New Zealand, can be seen as an example of the most up-to-date manifestation of inmate housing. This 232-bed, single-cell prison facility is a maximum security institution that incorporates all the amenities of a modern prison: a health center, education rooms, a gymnasium, dayrooms, laundry facilities, cafeteria, indoor/outdoor recreation centers, and wall-mounted stainless-steel furniture (sink/toilet, bunk beds, desk) for each cell. DLR Group's overall design is sleek, minimal, clean, and rather upbeat (if that can be said of a prison), with the individual cells looking as if they emerged from an ultra-modern *wallpaper** magazine spread or the advertising brochure of a hip Los Angeles hotel, rather than from the mind of criminal justice experts.

While DLR Group's design for Auckland Prison shows a marked increase of sympathy for the conditions of prisoners, the students of Archeworks, an "alternative design school" located in Chicago, felt that innovative solutions to prison architecture were still sorely lacking, so they created *Entropias* (a combination of the words "entropy" and "utopia"), a project which questions traditional prison designs and proposes radical alternatives to remedy the severe isolation of many prison structures. Their *Monitor* concept does away with the physical reality of an architectural space altogether, introducing a hybrid "virtual" architecture that prevents the criminal from repeating his or her illegal actions. The offender, in exchange for not being imprisoned, would either be forced to ingest a liquid tracking device or have a microchip implanted that would alert authorities of a potential crime through the monitoring of the criminal's vital signs. Trained specialists would then arrive to diffuse the situation. In the *Iris* design, the prison is replaced with a forced environmental facility dug deeply in the ground, covered by mesh and a tree canopy. *Iris* strives to rehabilitate inmates through the rehabilitation of the environment. Inmates farm, raise animals, and collect rainwater, and are forced to rely on each other in order to survive. Obviously, both *Entropias* projects are self-consciously utopian and their implementation seems highly unlikely, yet "real world" architects would do well to reinvestigate and reimagine housing strategies for convicted criminals.

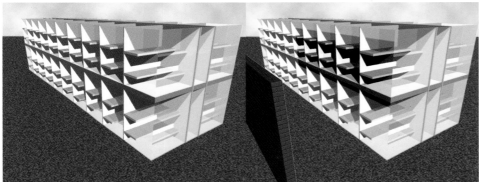

- The iris is a self contained, forced environmental, housing facility for convicted criminals.
  - It functions under minimal human surveillance and forces the inmates to rely on each other and their environmental setting for survival.
  - An inmate government will run the iris.
  - This system will not be their only life support, but in order make the best of their time they must learn to work within their new societal structure.
  - There will be no guards on the ground.
  - All surveillance will be made from the observation booth and prisoners will be followed with monitoring devices that will transmit their position.
  - The dome will be made of mesh, making the inmates deal with the outside conditions.
  - Inmates will farm, raise animals, and collect rainwater.
  - The program will be geared towards short to long-term criminals.
  - The tree canopy will be the wall between the inmates and outside society.

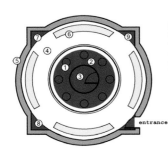

1. inmate seating
2. inmate circulation
3. officer area
4. visitor circulation
5. exterior
6. visitor seating
7. vending machines
8. restrooms
9. children area

Based on the Panopticon: Inmates in center separated from visitors-secure via corridor within another circle of security observation-as well as in visitor sphere

Without corridor there could be a central spiral stairway by which inmates and officers enter. Furthermore, the officer in the visitor sphere can make a full round

visitor center proposal by: Charlotte Dam

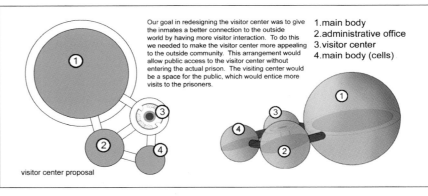

Our goal in redesigning the visitor center was to give the inmates a better connection to the outside world by having more visitor interaction. To do this we needed to make the visitor center more appealing to the outside community. This arrangement would allow public access to the visitor center without entering the actual prison. The visiting center would be a space for the public, which would entice more visits to the prisoners.

1. main body
2. administrative office
3. visitor center
4. main body (cells)

visitor center proposal

4-138 FRANCISCO TORRES

Isolation Room, Psychiatric Hospital of Cery
Lausanne, Switzerland, 2002

Francisco Torres, Lausanne, Switzerland

LEFT A RADICAL DEPARTURE FROM THE STERILE WHITE ROOMS NORMALLY ASSOCIATED WITH THE TREATMENT OF MENTAL ILLNESS, THIS ISOLATION CHAMBER HAS BEEN MADE MORE USER-FRIENDLY, EVEN MORE DOMESTIC, WITH CALMING COLORS, SOFT SHAPES, AND CONTEMPORARY DESIGN OBJECTS. THE TABLE IN THE CORNER ALSO FUNCTIONS AS A CHAIR WHEN LAID FLAT. THE PINKISH SCONCE ON THE WALL IS ACTUALLY A SILICONE-COVERED LAMP USED FOR LIGHT THERAPY.

Looking like the latest trend in minimal living, this pared-down bedroom is in fact a revolutionary intensive care unit designed by Francisco Torres in collaboration with the University Department of Adult Psychiatry in Vaud, Switzerland. Developed as part of Torres' master's degree in industrial design at the Ecoles spécialisées de la Suisse occidentale, the isolation chamber is intended to create a hypo-stimulant environment for patients suffering from aggression.

Our collective mental images of isolation rooms are often those conjured up by the media: sterile, white cells with harsh overhead lighting, padded walls, and lots of conspicuous restraining devices. And in reality these environments *are* angst-inducing if only because they are designed by engineers or hospital technical staff who tend to subjugate comfort to hygiene and safety. In an effort to change the image, Torres took a hands-on approach that involved spending time with patients and staff. By completing internships in several different units of the hospital, he familiarized himself with all of its different parts as well as the daily routines and health procedures of the various departments. Along the way, he was advised and monitored by a team of hospital staff.

The outcome of this collaborative effort is an atmosphere where hospital necessity meets household typology. The therapeutic room is sparse and clean but here and there touches of purple and green contribute to a feeling of calm and serenity, and bleached wood provides a bit of natural warmth. All accessories are practical and safe but also packaged in hip shapes and contemporary materials that mimic the domestic goods advertised in today's home improvement catalogues. A rubber-clad wall lamp, a table that doubles as an armchair, a silicone vase, and a floor mat cut from multi-colored pieces of plastic—who wouldn't want to overnight here? Torres even designed clothing for the patients, including reversible pants, T-shirts, and a pair of slippers with the names of great psychiatrists embossed on their soles.

The room as a whole is so strangely funky, it could almost be mistaken for a college dorm room, a furnished singles apartment, or the interior of the Big Brother house on page 142 of this book. And while well-designed environments have been proven to reduce anxiety and lower blood pressure, this radical room provokes many questions as to where—or whether—we draw the line between treatment and lifestyle.

4-140 LUCAS MAASSEN

Forever Young: Proportion Conflict, 2001
Eindhoven, The Netherlands

BOTTOM RIGHT AND OPPOSITE
WITH CHILDREN TALKING ON CELLULAR PHONES NOWADAYS, IT STILL IS ODD TO SEE A FOUR-YEAR-OLD IN A VERY ADULT KITCHEN THAT APPEARS TO HAVE BEEN SCALED DOWN TO FIT HER SIZE. IN ACTUALITY, THE KITCHEN WAS BUILT BY A LILLIPUTIAN WHO DESIGNED IT TO FIT HIS OWN NEEDS.

It doesn't take long to realize that something is awry in the series of photographs, *Forever Young,* by designer Lucas Maassen. Exactly what that "something" is has to do with what Maassen attributes to the "conflicts of proportion." Disturbed by what he saw as a surreal similarity between contemporary toddlers and adults, Maassen produced a study on proportional conflicts as his final academic project at the Design Academy in Eindhoven. His thesis was that children today act much older than their wee years, while adults strive to preserve their youth and act increasingly younger the older they get. For Maassen, the project was a study in how the behaviors of both intersect and the response of the material world to the phenomenon.

Forever Young is a visual expression of what happens when a four-year-old girl is placed in an adult kitchen. Upon careful observation of the cabinetry, garbage bins, pots and pans in this otherwise conventional environment it becomes apparent that the space couldn't possibly belong to an adult, but that the dimensions of the furniture have been compressed to fit a smaller body size. The scene is not faked, however, but a real kitchen owned, designed, and built by a Lilliputian who adapted the house for himself and his wife, so it was not necessary for Maassen to build a scaled-down set. For him, the existing model proved far more interesting because it had already been adapted to the needs of adults who are often treated as children because of their size. Maassen simply reversed the scenario—casting a child in the role of miniature adult—to picture the unnatural coalescence of "small" furniture and life-size appliances and, by extension, tiny tots and grown-ups.

With Hummers and Mercedes-Benzes now available for preteens, Maassen's study could not be more on-target. Pressured and cajoled by the media and consumer culture—which has an uncanny knack of reducing all ages to the common denominator of "twenty something"—it is increasingly obvious how much children and their parents are beginning to resemble one another. As critic Ralph Rugoff has pointed out, "In a culture which infantilizes everyone, there can be no generation gap. We all wear the same style clothing, listen to the same pop music, and wield the same remote to channel surf the same wavelengths."

Lucas Maassen, Eindhoven, The Netherlands

4-142 ENDEMOL

Big Brother House
Hilversum, The Netherlands, 1999–present

RIGHT AND OPPOSITE WHILE THE INTERIOR DESIGN COMES OFF AS TRENDY AND UPBEAT, LIFE INSIDE THE *BIG BROTHER HOUSE* DID NOT ALWAYS MATCH THE DECOR. RELENTLESSLY MONITORED BY CAMERAS LOCATED IN EVERY ROOM—ONE "GUEST" REMEMBERS HEARING THE LENS ZOOM IN ON HER—CONTESTANTS ON THE REALITY TV SHOW WERE NOT ALLOWED CONTACT WITH THE OUTSIDE WORLD.

Life inside the fish bowl takes on new significance in the context of reality TV shows such as *Big Brother*. Exactly 50 years after George Orwell released *Nineteen Eighty-Four*, a cautionary novel about life lived under 24-7 surveillance, a Dutch television production company made it a reality and sold the concept to 19 other countries, including Germany, England, and the United States.

Described as a real-life soap opera, the show revolved around a group of strangers who were locked up in a specially designed house rigged with television cameras and web cams that recorded their every move. The housemates themselves were completely sealed off from "reality" with no radio, TV, phone, Internet, or even writing materials. To fill the time, house members were given tasks by the producers of the show, via an unseen staff member referred to as "Big Brother," among which was the mandate to nominate fellow roommates for eviction. Every week the names of potential evictees were released during the show's broadcast so that viewers could vote for whom they wish to see removed.

Resembling the isolation tank on page 138 of this book, the "house" from the UK edition could better be described as a container in the likeness of a Habitat store, with a generous use of soothing purples and blues. Yet despite its trendy décor, the dwelling, like all BB residences, was very basic, with only the essentials of running water and a limited rationing of food. The minimal aesthetic, when combined with strict behavioral codes, round-the-clock monitoring, and human imperfection induced tension, uneasiness, and conflict—a precalculated design measure that assured viewer attention and brought in the big bucks. The formula proved so successful that some former contestants have been diagnosed with Post-Container Stress Disorder—a condition suffered by ex-members of the armed forces. One participant in Poland was actually taken to a psychiatric hospital.

Ironically, there was never any shortage of guinea pigs willing to place themselves under the scrutiny of the public eye for a relatively small payoff. Viewers, of course, never tire of human tragedies great and small, and in the Netherlands alone, 52 million web hits were recorded over the 100 days of the project.

Endemol, Hilversum, The Netherlands

4-144   MASAKI ENDOH & MASAHIRO IKEDA

Natural Ellipse, 2000-03
Shibuya, Tokyo, Japan

RIGHT Surrounded by more conventional buildings, Natural Ellipse looks like a giant white phallus or grain of rice. Its tructural cage, covered with a flexible skin, makes the building appear impenetrable.

OPPOSITE LEFT The third floor living quarters has a kitchen that wraps around the central stairwell. The all-white surfaces make the interior appear otherworldly.

OPPOSITE RIGHT At the building's apex a secret roof terrace is hidden, reached only through the fourth floor bathroom. The glass floor does double duty as a skylight.

Masaki Endoh and Masahiro Ikeda, Tokyo, Japan

The architects Masaki Endoh and Masahiro Ikeda are known for their "thin skin." Literally, this Japanese firm's signature style features buildings enveloped in taut, translucent, white skin that makes them look as if they are so light, they might just float away. One of their most recent projects, *Natural Ellipse*, is sited at the edge of Shibuya, Tokyo's shopping district and is surrounded by what the Japanese call "love hotels." Given its siting, some may humorously view the building as being vaguely phallic in shape, but the architects prefer to see it as a single grain of rice.

The overall geometry of the house is based on that of an elliptical toroid that was elongated at the same time it was being pierced at one end by a funnel, making the shape flare into a trumpet as it rises. While it appears to be a monolithic single-family dwelling, inside there are actually two apartments that can be entered from the street on opposite sides of the building. The first apartment occupies the first and second floors and has its own staircase, while the second apartment, the main one, revolves around the spiraling staircase which is the direct link between a basement study, entrance, and third-floor living quarters. The space is divided irregularly because of the geometry, with one side containing larger volumes—used for living and sleeping areas—and the other holding smaller spaces for kitchens and bathrooms. Windows have been placed at irregular intervals to capture certain views of the city. Rising gently though the core of the building is a delicate spiral staircase. Overall, materials and finishes are minimal and austere: painted steel plate, painted concrete floors, and painted mineral board walls that echo the aesthetic of the exterior.

At the building's apex is a surprise: a secret terrace hidden by the steeply indented walls. The terrace's crown jewel is a glass floor that also acts as a skylight, filtering light through the interior via the stairwell.

4-146

Masaki Endoh and Masahiro Ikeda, Tokyo, Japan

LEFT AT THE CENTER OF THE BUILDING, STRETCHING FROM BASEMENT TO THE THIRD FLOOR IS A DELICATE SPIRALING STAIRCASE. ITS PERFORATED METAL STEPS ALLOW LIGHT TO FILTER THROUGH THE WHOLE BUILDING.

4-148 CHRIS BURDEN

with TK Architecture, Southern California, USA
small skyscraper, 2001
developed as part of the Houses x Artists project (1998–2002)

RIGHT AND OPPOSITE

FRUSTRATED BY LA COUNTY BUILDING CODES, ARTIST CHRIS BURDEN SKETCHED THE FIRST DRAWING OF SMALL SKYSCRAPER IN 1994—HERE SEEN, AS IT WOULD APPEAR ON HIS PROPERTY.
IN EARLIER DRAWINGS, A SPIRAL STAIRCASE CONNECTED THE FLOORS, COSTING PRECIOUS AMOUNTS OF SPACE TO BE WASTED. LATER, AN ELEVATOR REPLACED THE STAIRCASE.
"IT'S KIND OF LIKE A MODERN DAY LOG CABIN," SAYS BURDEN. HERE, A CAD DRAWING OF THE SMALL SKYSCRAPER.
AN INSTALLATION VIEW OF THE FULL-SCALE PROTOTYPE OF SMALL SKYSCRAPER. BURDEN USED ALUMINUM AND WOOD TO BUILD A FEATHERWEIGHT STRUCTURE THAT COULD BE CONSTRUCTED BY UNTRAINED BUILDERS WITH A MINIMUM OF TOOLS.

For an artist like Chris Burden, who began his notorious career by confining himself for days to school lockers or sitting for weeks on end in a gallery, hidden on a platform pressed close to the ceiling, the possibility of living in a structure that encompasses 400 square feet and is 35 feet tall, must feel like a generous proposition. For most of us, though, it might feel daunting to live within the confines of the *small skyscraper* project proposed by Burden and developed with TK Architecture for the Houses x Artists project (pronounced "Houses by Artists").

In 1998, Alan Koch and Linda Taalman of TK Architecture (formerly of openoffice), invited nine artists, many of whom have architectural tendencies, to co-design with them a house with the single stipulation being that their ideas, however conceptual, result in a house that could be built. For Burden, his *small skyscraper* project became an experiment in living within limitations. He found a loophole in the Los Angeles county building code that allows individuals to build small outbuildings—like greenhouses and sheds—without first obtaining permits if the structure encompasses no more than 400 square feet and is no more than 35 feet tall. Taking this loophole as creative license, the artist/architect team designed a four-story structure where each floor measures a mere ten by ten feet. Each floor is approximately eight feet high with a portion of each floor reserved for a one-person solar-powered elevator. All the main rooms of a traditional house or apartment are in evidence, just in a compressed form. The first floor features an entryway and abbreviated living room; the second floor, a compact kitchen and dining area, the third floor houses the bathroom, and the bedroom is nestled on the fourth floor. There is even access to a semi-legal rooftop terrace.

Imagined as a retreat or "like a modern-day log cabin," Burden's *small skyscraper*, currently still in prototype form (realized by Los Angeles contemporary Exhibitions), is designed as a featherweight structure that can be constructed from a kit of aluminum parts. Wood floors are made up of loosely laid 2x4s that can be easily removed or rearranged and a glass-cladding system with sliding doors at the corners of each room. The kit can be erected by untrained builders with a minimum of tools and equipment at a fraction of the cost of a more traditional domestic dwelling. Burden plans on erecting the *small skyscraper* on his property in Topanga Canyon in Los Angeles in the near future.

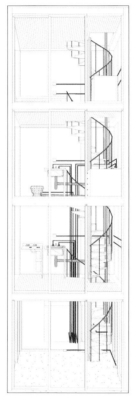

chris burden, Los Angeles, California, USA

4-150 ABSALON

Paris, France
Cellule No. 1 (Réalization Habitable), 1993

Cellule No. 1 is the first in a series of six dwellings that represent the culmination of one man's quest to achieve peace, tranquility, and perfect harmony through clean lines, perfect scale, and clarity of form. One of the last major works of Israeli-born artist Absalon, who died in 1993 at the age of 29, *Cellule No. 1* is a self-contained urban survival unit constructed in Paris and in which the artist intended to live.

Comprised of two simple geometric shapes, a rectangle and a cylinder, the oblong form is punctuated by long horizontal openings (inspired by Le Corbusier) and one entry portal. All walls, inside and out, are painted white and interior lighting comes from a neon tube. Decoration and superfluous detail are omitted, all living supplies reduced to the bare necessities: a horizontal board for sleeping, a simple bench for eating, and an upright basin for washing.

Scaled almost exactly to the dimensions of his own body, *Cellule No. 1* is constraining, suffocating, and alienating. It is difficult to look at videos of it in use without recalling the tension-filled straightjacket escapes of twentieth-century magician Houdini. At the same time, the *Cellule* is also a sanctuary and retreat from the impurities of the real world, a serene and meditative place that demands of its user the holier traits of self-discipline and patience. Like the hermetic "safe rooms" for sufferers of allergies, made famous in the Todd Haynes film, *Safe* (1995), the pod isolates and protects.

A precisionist with a monk's stamina, Absalon subjected himself willingly to the rigors of the cell in order to restructure his own personal behavior, envisioning "not public projection or the improvement of the human condition, but the intellectual and spiritual intensification of the private domain."

RIGHT EXTERIOR VIEW OF *CELLULE NO. 1*, ONE IN A SERIES OF SIX PRISTINE WHITE CELLS DESIGNED BY ISRAELI-BORN ARTIST ABSALON. THE *CELLULES* WERE INTENDED TO BE PLACED IN SIX DIFFERENT URBAN SETTINGS, INCLUDING TEL AVIV, FRANKFURT, AND ZURICH. ONLY TWO WERE ACTUALLY REALIZED, THIS ONE IN PARIS.

OPPOSITE ONCE INSIDE THE *CELLULE*, THE USER'S PERSONAL BEHAVIOR IS ALTERED BY THE CRAMPED AND PRISTINE SPACES.

Absalon, Paris, France

4–152 GREGOR SCHNEIDER

Rheydt, Germany
Totes Haus ur (Dead House ur), 1985–present

Gregor Schneider, Rheydt, Germany

LEFT GERMAN ARTIST SCHNEIDER HAS REWORKED THE INTERIOR OF HIS HOME SO MANY TIMES THAT HE CAN NO LONGER REMEMBER WHAT THE ORIGINAL LAYOUT LOOKS LIKE.

BELOW RIGHT THE SOUND-PROOFED BASEMENT.

since the age of 16, german artist gregor schneider has been compulsively transforming the house where he lives and works in Rheydt, a small town outside of cologne, into an archaeological wonder. seen from the street, the house is a relatively straightforward example of the austere architecture of the industrial Rhine region. Inside, however, is a wholly different slice of domestic strata.

what appear to be banal middle-class rooms are, in fact, duplicates of the rooms in which they are situated, slightly scaled-down and painstakingly crafted to resemble the originals. walls have been inserted in front of existing walls, windows in front of windows, doors in front of doors. some rooms have several windows arranged one behind another with lamps and ventilators in between to simulate daylight and a gentle wind. The basement has been reworked into an oppressive cell and its walls lined with sound-insulating lead and plastered over. sometimes visitors are invited over for "coffee and cake" in the homely dining room, an eerie sliver of its former self that revolves on its own axis, with imperceptible slowness, thanks to a mechanical turntable underneath. with each action schneider takes, rooms and corridors become smaller, openings close (or open somewhere else), paths become more labyrinthine, and the house slowly implodes. with the passage of time, a linear sense of time disappears, and the artist admits he can no longer remember or reconstruct the home's original layout.

unlike other homes, which change slowly—if ever—over time, allowing memories to be formed and savored in due course, schneider's house is a frenetic and continuously moving organism that exists in a state of perpetual flux. the artist's own reminiscences take the form of photos and videos, which he makes after completing a set of alterations. straight-on photographs in black and white, or home movies made by stumbling through the house with a hand-held camera, provide a stunning psychological portrait of a house under pressure.

4-154 AL V. CORBI, THE DESIGNER
Safe Homes, 1999–present
Los Angeles, California, USA

RIGHT Situated on a hillside in Los Angeles is the 11,000 square foot mansion of security guru Al Corbi built from 1,600 yards of concrete (16 times more than used in a typical house) and protected by a nine-million-dollar security system.

OPPOSITE As Corbi is quick to point out, it takes on average seven to ten seconds for an intruder to make their way from the break-in point to the master bedroom. Unfortunately, it takes most people 15 seconds to wake up, which means fail-proof security is an absolute must. He uses 650-pound Italian-made doors with chromium steel bolts.

FOLLOWING PAGES Banal interiors hide the security systems imbedded within them. The mattress in the master bedroom, for example, sits above a double-layer of bullet-proof Kevlar, should anyone want to shoot at Corbi or his wife from the dining room below. A variety of household monitors are ready to go into "event mode" and display an intruder's activities while also warning the family of the imminent danger.

Paranoia can be the mother of invention. At least it has been for Los Angeles security expert Al Corbi, who has turned the nation's fear of imminent attack into a million dollar business. Fueled by the attacks of September 11, 2001, news reports of violent crime, and films such as *Panic Room*, Americans have never felt more vulnerable.

Corbi, who worked for the Department of Justice for 23 years creating highly classified secure offices, recommends an encompassing plan of defense. His own five-story, 11,000 square-foot home is equipped with nine million dollars worth of visible security cameras, an undisclosed number of smart security cameras, 36 telephones interconnected with an intercom pager system, hidden weapons systems, and armor-piercing walls. Dispersed throughout the house are weapons, shields, and warning devices that can stave off everything from burglary and kidnapping to missile, rocket, and biological attacks.

Disguise is key, with monitors, motion detectors, and defense systems integrated into otherwise typical interiors. Should any would-be prowler actually make it past the outdoor surveillance, he or she would be stopped dead in their tracks, quite literally, by a spray of spring-loaded spikes tucked under the floorboards in front of Corbi's master bedroom. Should this fail, a kill-zone system of short-barreled shotguns in the walls can open fire when triggered from a remote control. The bed itself is protected by a double layer of bullet-proof Kevlar, and the entire room is a ballistics-grade box—or safe core—that can be sealed off in seconds, isolating the bad guys or poisonous chemicals outside. In the unlikely event that all else fails, a vertical transporter is ready to whisk Corbi and family deep within the hill their home is situated on, into an escape tunnel below.

Indeed it is the language of fear that convinces Corbi's clients to invest in similar products. Who could ignore Corbi's warnings? "Once our alarms are triggered, it takes at least 30 minutes for help to arrive in most cities, longer, if ever, in others…since the average theft and violent crime is concluded in 15 minutes or less, this leaves the police with little more to do than count bodies or take a report." Reiterating an opinion shared by many Americans in the recently released and highly controversial gun-control documentary, *Bowling for Columbine*, Corbi proclaims: "We can no longer rely on the highly questionable ability of others to protect us and our families. We must do this most important task ourselves."

4-156

al v. corbi, Los Angeles, California, USA

BIBLIOGRAPHY

CHAPTER 1
SOME ASSEMBLY REQUIRED

Dwell, February 2002 (article on Ken Draizen).
Gloger, Katja, "IKEA: Ein Mann vermöbelt die Welt," *Stern*, no. 18, April 24, 2003, pp. 82–92.
Gluck, Robert, "Fee, Fi, Faux, …Fum," *Nest*, winter 2001–02, pp. 68–81.
Maffeo, Lois, "Fort Thunder," *Nest*, summer 2001, pp. 134–55.
Moody, Tom, "Charles Stagg," webblog, April 11, 2002, (www.digitalmediatree.com/tommoody).
Oppenheimer Dean, Andrea and Timothy Hursley, *Rural Studio: Samuel Mockbee and an Architecture of Decency*, New York: Princeton, 2002.
Ray, Oliver, "The Workhouse," *Nest*, winter 2002–03, pp. 23–26.
Stadler, Matthew, "The Raw and the Cooked," *Nest*, fall 2001, pp. 180–93.
Tlali, Miriam, "Mrs. Zama's House," *Nest*, no. 15, winter 2001–02, pp. 152–61.

CHAPTER 2
MOVING PICTURES

Cameron, Kristi, "Pocket Penthouse," *Metropolis*, February 2003, pp. 86–87.
Chatterjee, Pratap, "Force Provider: The Base in a Box," *CorpWatch*, May 2, 2002.
Hulsman, Bernard, "Box of Tricks," *World Architecture*, November–December 2001.
Lehmann, Ulrike, "Malerie in der Fotografie," *Kunstforum*, vol. 164, March–May 2003. November 3–December 22, 2001.
"Preston Scott Cohen: Toroidal Architecture," press release, Thomas Erben Gallery, New York, November 3–December 22, 2001.
Riley, Terence, *The Un-Private House*, New York: MoMA, 1999.
sehrgut.de/mib/hans_e.php: "Made in Berlin."
Skoggard, Carl, "An Immaculate Conception," *Nest*, spring 2002.
talkabout.it, "Maurice Agis," April 2000.
Topham, Sean, "Shock of the Pneu," *Blowup*, Munich/Berlin/London/New York: Prestel, 2002, pp. 90–91.
"The TransHab Module: An Inflatable Home in Space," *NASA Facts*, Lyndon B. Johnson Space Center, May 1999.
www.abcnews.com, "Soldier of the Future: With New Technology, He Might Fight Like Robocop, Drive Like James Bond."
Young, Paul, "Keep Clear," *Surface*, issue 40, p. 84.

CHAPTER 3
GO WITH THE FLOW

Hamilton, William L., "For Home Theaters, Chairs are a Blockbuster," *New York Times*, May 22, 2003.
Searle, Adrian, "Have You Ironed Your Room Yet?," *The Guardian*, April 23, 2002.
www.kempert.org/exhibits: Self, Dana, "The Perfect Home."
Such, Robert, "Heretical Tent," *Architecture Week*, June 19, 2002, p. E2.2.

CHAPTER 4
CONTENTS UNDER PRESURE

Bruthansová, Tereza, "Matali Crasset," *Blok* (www.blok-online.cz/04/0404en.htm).
"Entropias," Archeworks publication, 2000–01.
Gitlen, Laurel and Annette Ferrara, "Bringing the Dream Home: Four Living Spaces," *tenbyten*, vol. 2, no. 1, spring/summer 2002, pp. 38–41.
Gleason, Joshua, "Imprisoning Form," *tenbyten*, vol. 1, no. 3, fall 2001.
Jeffett, William, "Atelier Van Lieshout," *NYArts* magazine, 1999.
McGuire, Penny, "Eastern Pleasure," *Architectural Review*, April 2003, pp. 74–77.
Riemschneider, Burkhard and Uta Grosenick, *Art at the Turn of the Millennium*, Cologne: Taschen, 1999, pp. 14 and 450.
Rooijen, Jeroen van, "Moving Up in the World," *Frame*, March/April 2003, pp. 130–31.
Rugoff, Ralph, "Children and Other Miniature Collectibles," *Circus Americanus*, London: Verso, 1995, pp. 145–47.
Schouwenberg, Louise, "Forever Young," *Frame*, March/April 2003, pp. 122–23.
Torres, Francisco, "Creation of an Intensive Care Room for a Psychiatric Hospital," October 2002.
www.galleriesatmoore.org/publications/absalonpv: Vergne, Philippe, "Absalon: The Man Without a Home Is a Potential Criminal."

WEB & E-MAIL ADDRESSES

Maurice Agis	www.dreamspace-agis.com
Atelier van Lieshout	www.avl-ville.com
Berkline, Inc.	www.berkline.com
Big Brother	www.wikipedia.org/wiki/Big_Brother_television_program
The Designer (Al v. Corbi)	www.thedesignerusa.com
DLR Group Architecture	www.dlrgroup.com
Dornbracht	www.dornbracht.de
Marion Eichmann	www.marioneichmann.com
Endemol	www.endemol.nl
FAT	www.fat.co.uk
Hans Hemmert	hans.hemmert@berlin.de
IKEA	www.ikea.com
Aleksandra Konopek	office@konopek.de
LOT/EK	www.lot-ek.com
Lucas Maassen	lucas@unitedstatements.nl
NASA	www.nasa.gov
pool Architektur	pool.helma.at
Rural Studio	www.ruralstudio.com
Gregor Schneider	office@konradfischergalerie.de
Seoungwon Won	firstwon72@hotmail.com
Snowcrash	www.snowcrash.se
Sod Houses	www.websteader.com
Softroom	www.softroom.com
Tsui Design & Research	www.tdrinc.com
Stefan Wischnewski	wischnewski@gmx.net

PHOTO CREDITS

All reasonable efforts have been made to obtain copyright permission for the images in this book. If we have committed an oversight, we will be pleased to rectify it in a subsequent edition.

Front jacket: courtesy carlier | gebauer
Frontispiece: copyright by Luuk Kramer *fotografie*
p. 9, left: Kunsthalle Tübingen, collection of Prof. Dr. Georg Zündel/VG-Bildkunst.
p. 13, clockwise from top left: © Hiroyuki Hirai; © Bodyshower®; photo by Rick Miller; © InterAMI Interior, Ukraine.
p. 11, top: © Barbara Gallucci 2002; © Michael Sailstorfer (with Alfred Kurz).
p. 15, clockwise from top left: Wasmuths Monatshefte für Baukunst; courtesy of the Art Institute of Chicago; photo by Wolfgang Thöner, 2002; © Robert Bruno; © ARS 2003 and © VG-Bildkunst 2003.
p. 17, clockwise from top left: © Warner Brothers; Archive Fundación César Manrique, Lanzarote; © Verner Panton Design; photo Michel Moch, Paris; Archigram Archives; © MGM.
pp. 22–25: © Hisham Akira Bharoocha
pp. 26–27: © Inter IKEA Systems B.V.
pp. 28–31: Zwelethu Mthethwa
pp. 32–33: David Wood © 2003. All rights reserved DLW Graphics/websteader.com
pp. 34–35: Noah David Smith
pp. 36–39: courtesy LOT/EK. Photos © Paul Warchol Photography Inc.
pp. 40–43: © Timothy Hursley
pp. 44–47: photos by Janna and John Fulbright, John Fulbright Studios
pp. 48–49: © Ejlat Feuer
pp. 50–53: courtesy Eugene Tsui
pp. 58–61: courtesy Stefan Wischnewski
pp: 62–65 © Hartmann and Wetzel
pp. 66–67: www.snowcrash.se. Photos by Bengt O. Pettersson
pp. 68–69: www.snowcrash.se. Photos by Bobo Olsson
pp. 70–71: courtesy of the artist and Lehmann Maupin, New York
pp. 72–75: © R & Sie...
pp. 76–79: © NASA
pp. 80–81: courtesy Thomas Erben Gallery, New York
pp. 82–83: © 2003 VG-Bildkunst, Bonn
pp. 84–85: © Berkline, Inc.
pp. 86–89: photos by Uwe Spoering, © Dornbracht
pp. 94–97: courtesy carlier | gebauer
pp. 98–99: courtesy Marion Eichmann, photos by Harry Schnitger and Gerd Engelsmann
pp. 100–102: photos by Hertha Hurnaus
pp. 104–105: U.S. Army
pp. 106–109: Softroom
pp. 110–113: copyright by Luuk Kramer *fotografie*
pp. 114–115: © Ennemlaghi
pp. 116–119: © Carolyn Schaefer/VEGAMG
pp. 120–123: © Seoungwon Won
pp. 124–125: courtesy Maurice Agis. Photos (left) by D Lauder; (right) by P Brotons
pp. 130–131: FAT
pp. 132–135: courtesy Atelier van Lieshout. Photo of *sleep skull* by AVL; *Tampa skull* by Peter Foe. © VG-Bildkunst, Bonn 2003
p: 136: © DLR Group
p. 137: © Ammar Eloueini, Louis Schalk, Francis Kmiecik
pp. 138–139: photos by Anoush Abrar/ECAL
pp. 140–141: © Lucas Maassen
pp. 142–143: © Channel 4
pp. 144–147: photos by Hiroyasu Sakaguchi
pp. 148–149: photos courtesy of TK Architecture
pp. 150–151: courtesy Galerie Chantal Crousel
pp. 152–153: © Gregor Schneider
pp. 154–157: © Al V. Corbi

THE AUTHORS AND PUBLISHER ARE VERY GRATEFUL TO THE FOLLOWING FOR THEIR HELP IN THE CREATION OF THIS BOOK:

MAURICE AGIS; ARCHEWORKS; HISHAM BHAROOCHA; STEPHEN BRAM; TIM BROWN; THE CHURCH OF CRAFT; AL AND LANA CORBI; DO-HO SUH; RENO DAKOTA; ISABEL DOTZAUER, LOTHRINGER DREIZEHN; KEN DRAIZEN; STEFAN EBERSTADT; MARION EICHMANN; MELISSA FOSTER DENNEY, RURAL STUDIO PROGRAM SPECIALIST, AUBURN UNIVERSITY; AMMAR ELOUEINI; ENNEMLAGHI; LANA FULBRIGHT; BARBARA GALLUCCI; JAMES GIBBS; LAUREL GITLEN; JOSHUA GLEASON; DON GUSS; URS HARTMANN AND MARKUS WETZEL; HANS HEMMERT; MARJAN HOLMER; CURT HOLTZ; PHILIPPA HURD; IKEA; SAM JACOBS, FAT; GALERIE CAROL JOHNSSEN, MUNICH; PROFESSOR TÖNIS KÄO, INSTITUTE DESIGN IN RESEARCH, WUPPERTAL; ALEKSANDRA KONOPEK; JENNIFER LIESE; ATELIER VAN LIESHOUT; GIUSEPPE LIGNANO, LOT/EK; LUCAS MAASSEN; TOM MOODY; ALEXANDRA RAMSEY; SIMONE SCHMICKL; ELSA DE SEYNES, CARLIER GEBAUER; RURAL STUDIO; STUART SMITH + MANI SURI (SMITH); SOFTROOM; CHARLES STAGG; JAMES MICHAEL TATE; *TENBYTEN;* TK ARCHITECTURE; SEAN TOPHAM; FRANCISCO TORRES; EUGENE TSUI; PATTY WELSH; STEFAN WISCHNEWSKI; FLORIAN WALLNÖFER; SEOUNGWON WON